Neuroscience, Neurophysiology and Symmetry

Neuroscience, Neurophysiology and Symmetry

Editors

Thierry Paillard
Sandeep Kumar Singh

MDPI • Basel • Beijing • Wuhan • Barcelona • Belgrade • Manchester • Tokyo • Cluj • Tianjin

Editors
Thierry Paillard
Université de Pau et des Pays de l'Adour
France

Sandeep Kumar Singh
Indian Scientific Education and Technology (ISET) Foundation
India

Editorial Office
MDPI
St. Alban-Anlage 66
4052 Basel, Switzerland

This is a reprint of articles from the Special Issue published online in the open access journal *Symmetry* (ISSN 2073-8994) (available at: https://www.mdpi.com/journal/symmetry/special_issues/Neuroscience_Neurophysiology_Symmetry).

For citation purposes, cite each article independently as indicated on the article page online and as indicated below:

LastName, A.A.; LastName, B.B.; LastName, C.C. Article Title. *Journal Name* **Year**, *Volume Number*, Page Range.

ISBN 978-3-0365-7478-3 (Hbk)
ISBN 978-3-0365-7479-0 (PDF)

Contents

About the Editors

Thierry Paillard

Thierry Paillard is a full professor at University of Pau et Pays de l'Adour, France. He holds a PhD in sports science (exercise physiology) and a PhD in neuroscience (neurophysiology). In general, his research focuses on movement and postural balance. More specifically, his research themes are the plasticity of postural and motor function and electrophysiology. He is currently director of the Movement, Balance, Performance and Health laboratory (EA 4445).

Sandeep Kumar Singh

Sandeep Kumar Singh (PhD) is Research Scientist at the Indian Scientific Education and Technology Foundation, India. He holds a PhD Degree in Biotechnology. He has more than 10 year research experience in Neuroscience. In general his research focus on Therapeutic aspects of Alzheimer's disease, focus on use of natural compounds and nutraceuticals for the treatment of Alzheimer's disease.

Preface to "Neuroscience, Neurophysiology and Symmetry"

Human movements and posture often show lateral asymmetries in healthy, young, older, frail and pathological subjects. Why are there asymmetries in motor behaviour? These asymmetries have not been fully identified and are likely to stem from different components of motor function, such as the sensory (perception), central (central integration) and motor (movement command and control) components. The neural mechanisms involved are also not yet understood at different neurological levels (peripheral, spinal, subcortical and cortical). Therefore, exploratory research is needed in order to understand symmetry or asymmetry in terms of human movement and posture. This Special Issue, "Neuroscience, Neurophysiology and Symmetry", presents experimental or theoretical data that provide answers to these questions, focusing mainly on asymmetry in human movement and posture.

Thierry Paillard and Sandeep Kumar Singh
Editors

symmetry MDPI

Editorial

Asymmetry of Movement and Postural Balance and Underlying Functions in Humans

Thierry Paillard

Laboratory of Movement, Balance, Performance and Health (MEPS), University of Pau and Pays de l'Adour, E2S, 65000 Tarbes, France; thierry.paillard@univ-pau.fr

Citation: Paillard, T. Asymmetry of Movement and Postural Balance and Underlying Functions in Humans. *Symmetry* **2023**, *15*, 759. https://doi.org/10.3390/sym15030759

Received: 14 March 2023
Accepted: 16 March 2023
Published: 20 March 2023

Human movements and posture often show lateral asymmetries. Although symmetry is not systematically observed between two limbs, its presence is likely to influence motor and postural performance and the risk of injury and falls in sportspeople, healthy, elderly, and frail subjects during professional, sports and leisure activities, as well as activities in day to day life [1–3]. A systematic search for possible inter-limb asymmetry in the context of the optimization of motor performance or the rehabilitation of functional abilities can be undertaken. Inter-limb symmetry or asymmetry may occur as a function of motor experience (e.g., high versus low), the nature of movements (e.g., specialized versus non-specialized), the environmental context (e.g., easy versus difficult motor tasks), individual or intrinsic factors (e.g., proprioception, hemispheric laterality, motor output) and the limb dominance effect. The finer details of motor and postural symmetry or asymmetry have not yet been fully identified in terms of information perception, central integration and movement command and control [4]. In addition, the neural mechanisms involved are also not fully understood at the different neurological levels (peripheral, spinal, subcortical and cortical). Therefore, exploratory research is needed in order to understand symmetry or asymmetry in terms of human movement and posture. This Special Issue, "Neurosciences, Neurophysiology and Symmetry", includes six papers that provide some answers to these questions, focusing mainly on asymmetry in human movement and posture.

The first paper, by Barbara Dobies-Krześniak et al. [5], tests the hypothesis that functional laterality features are associated with scoliosis incidence (radiologically confirmed as idiopathic scoliosis). Side dominance was determined by the lateral preference inventory. The direction, strength and consistency of lateral dominance were evaluated. Lateralization analysis showed some trends, but the results obtained were not statistically significant. Thus, the relationship between scoliosis and laterality may not be a simple causal relationship, and needs further investigation.

The second paper is based on the kinematic analysis of lower limb joint asymmetry during gait in people with multiple sclerosis [6]. The majority of people with multiple sclerosis (pwMS), report lower limb motor dysfunctions, which may affect postural control, gait and a wide range of daily activities. While it is quite common to observe a differing impact of the disease on the two limbs (i.e., one of them can be more affected), less clear are the effects of such asymmetry on gait performance (kinematic and spatio-temporal parameters with eight-camera motion capture system). Based on cyclogram orientation and trend symmetry, the results showed that pwMS exhibit significantly greater asymmetry in all three joints (hip, knee, ankle) than unaffected individuals. Moreover, the same parameters were sensitive enough to discriminate individuals of different disability levels. With few exceptions, all the calculated symmetry parameters were found to be significantly correlated with the main spatio-temporal parameters of gait and the EDSS (Expanded Disability Status Scale) score. In particular, large correlations were detected between Trend Symmetry and gait speed and between trend symmetry and EDSS score. Such results suggest not only that MS is associated with significantly marked interlimb asymmetry during gait, but also that such asymmetry worsens as the disease progresses and that it has a relevant impact on gait performances.

Based on the fact that the current literature shows no consensus regarding the difference between the dominant leg (D-Leg) and the non-dominant leg (ND-Leg) in terms of postural control, the third paper deals with the effects of limb dominance on postural balance in sportsmen practicing symmetric and asymmetric sports [7], the hypothesis being that the lack of consensus could stem from motor experience (i.e., symmetric or asymmetric motricity) and/or the physiological state induced by physical exercise. The study by Kadri et al. [7] aimed to investigate the acute effects of fatiguing exercise on postural control when standing on the D-Leg and the ND-Leg in athletes practicing symmetric (SYM) and asymmetric (ASYM) sports. Monopedal postural control was assessed for the D-Leg and the ND-Leg before and after the fatigue period (which consisted of repeating squats until exhaustion). A force platform was used to calculate the spatio-temporal characteristics of the displacements of the center of foot pressure (COP). A significant fatigue effect was observed in both groups on the D-Leg and the ND-Leg for all the COP parameters. There was a strong tendency ($p = 0.06$) between the ASYM and SYM groups on the D-Leg, concerning the relative increase in the COP velocity in the frontal plane after the fatigue period. The fatigue condition disturbed postural control in both the SYM and ASYM groups on the D-Leg and ND-Leg. This disturbing effect related to fatigue tends to be more marked in athletes practicing asymmetric sports than in athletes practicing symmetric sports on the D-Leg. Effects related the nature of sports (SYM and ASYM) and muscle fatigue need to be confirmed (or negated) by future studies.

The fourth paper is innovative and is entitled "From Neural Command to Robotic Use: The Role of Symmetry/Asymmetry in Postural and Locomotor Activities" [8]. This article deepens a reflection on why and how symmetry/asymmetry affects the motor and postural behavior from the neural source, through uterine development and child maturation, and how the notion of symmetry/asymmetry has been applied to the design and control of walking robots. The concepts of morphology and tensegrity are also presented, to illustrate how the biological structures have been used in both sciences and arts. The development of the brain and the neuro-fascia-musculoskeletal system seems to be relatively symmetric from the beginning of life through to complete maturity. The neural sources of movements (i.e., central pattern generators) are able to produce either symmetric or asymmetric responses to accommodate environmental constraints and task requirements. Despite the fact that human development is mainly symmetric, asymmetries regulate neurological and physiological development. Laterality and sports training could affect the natural musculoskeletal symmetry. The plasticity and flexibility of the nervous system allows it to adapt and compensate for environmental constraints and musculoskeletal asymmetries in order to optimize the postural and movement control. For designing humanoid walking robots, symmetry approaches have been mainly used to reduce the complexity of the online calculation. Applications in neurological retraining and rehabilitation should also be considered.

The following two articles do not deal specifically with motor and postural asymmetries, but focus more broadly on cerebral asymmetries and functional abilities. The paper by Van Vuuren et al. [9] is based on the fact that neonatal and adult strokes are more common in the left than in the right cerebral hemisphere in the middle cerebral arterial territory, and adult extracranial and intracranial vessels are systematically left-dominant. The aim of the research reported here was to determine whether the asymmetric vascular ground plan found in adults was present in healthy term neonates. A new transcranial Doppler ultrasonography dual-view scanning protocol, with concurrent B-flow and pulsed wave imaging, acquired multivariate data on neonatal middle cerebral arterial structure and function. This study documents systematic asymmetries in the middle cerebral artery origin and distal trunk of healthy term neonates for the first time, and identifies commensurately asymmetric hemodynamic vulnerabilities. A systematic leftward arterial dominance was found in the arterial caliber and cortically directed blood flow. The endothelial wall shear stress was also asymmetric across the midline, and varied according to vessels' geometry. They conclude that the arterial structure and blood supply in the brain are laterally asymmetric in newborns. Unfavorable shearing forces, which are a by-product of the arterial asymmetries described here, might contribute to a greater risk of cerebrovascular pathology in the left hemisphere.

The last paper dealt with the neurological asymmetry of self-face recognition [10]. While the desire to uncover the neural correlates of consciousness has taken numerous directions, self-face recognition has been a constant in attempts to isolate aspects of self-awareness. The neuroimaging revolution of the 1990s brought about systematic attempts to isolate the underlying neural basis of self-face recognition. These studies, including some of the first fMRI (functional magnetic resonance imaging) examinations, revealed a right-hemisphere bias for self-face recognition in a diverse set of regions, including the insula, the dorsal frontal lobe, the temporal parietal junction, and the medial temporal cortex. This systematic review provides confirmation of these data (which are correlational), which were provided by TMS (transcranial magnetic stimulation) and patients in which the direct inhibition or ablation of right-hemisphere regions leads to a disruption or absence of self-face recognition. These data are consistent with a number of theories, including a right-hemisphere dominance for self-awareness and/or a right-hemisphere specialization for identifying significant social relationships, including to oneself.

In conclusion, asymmetries of motor and postural functions and many other organic functions in humans are likely to be observed at any age, in healthy and pathological subjects, and deserve to be fully explored in order to better anticipate and (possibly) prevent them. In this ambition, scientific research in this field is still in its infancy.

Conflicts of Interest: The author declares no conflict of interest.

References

1. Bishop, C.; Turner, A.; Read, P. Effects of inter-limb asymmetries on physical and sports performance: A systematic review. *J. Sports Sci.* **2018**, *36*, 1135–1144. [CrossRef] [PubMed]
2. Brown, S.R.; Brughelli, M.; Lenetsky, S. Profiling Single-Leg Balance by Leg Preference and Position in Rugby Union Athletes. *Motor Control* **2018**, *22*, 183–198. [CrossRef] [PubMed]
3. Laroche, D.P.; Cook, S.B.; Mackala, K. Strength asymmetry increases gait asymmetry and variability in older women. *Med. Sci. Sports Exerc.* **2012**, *44*, 2172–2181. [CrossRef]
4. Paillard, T.; Noé, F. Does monopedal postural balance differ between the dominant leg and the non-dominant leg? A review. *Hum. Mov. Sci.* **2020**, *12*, 74. [CrossRef]
5. Dobies-Krześniak, B.; Werblińska, A.; Tarnacka, B. Lateralization Direction, Strength, and Consistency in Juvenile and Adolescent Idiopathic Scoliosis: A Case Control Pilot Study. *Symmetry* **2022**, *14*, 888. [CrossRef]
6. Pau, M.; Leban, B.; Deidda, M.; Putzolu, F.; Porta, M.; Coghe, G.; Cocco, E. Kinematic Analysis of Lower Limb Joint Asymmetry During Gait in People with Multiple Sclerosis. *Symmetry* **2021**, *13*, 598. [CrossRef]
7. Kadri, M.A.; Noé, F.; Maitre, J.; Maffulli, N.; Paillard, T. Effects of Limb Dominance on Postural Balance in Sportsmen Practicing Symmetric and Asymmetric Sports: A Pilot Study. *Symmetry* **2021**, *13*, 2199. [CrossRef]
8. Blanchet, B.; Guertin, P.; Pilon, F.; Gorce, P.; Prince, F. From Neural Command to Robotic Use: The Role of Symmetry/Asymmetry in Postural and Locomotor Activities. *Symmetry* **2021**, *13*, 1773. [CrossRef]
9. Van Vuuren, A.J.; Saling, M.; Rogerson, S.; Anderson, P.; Cheong, J.; Solms, M. Cerebral Arterial Asymmetries in the Neonate: Insight into thePathogenesis of Stroke. *Symmetry* **2022**, *14*, 456. [CrossRef]
10. Janowska, A.; Balugas, B.; Pardillo, M.; Mistretta, V.; Chavarria, K.; Brenya, B.; Shelansky, T.; Martinez, V.; Pagano, K.; Ahmad, N.; et al. The Neurological Asymmetry of Self-Face Recognition. *Symmetry* **2021**, *13*, 1135. [CrossRef]

symmetry

MDPI

Article

Lateralization Direction, Strength, and Consistency in Juvenile and Adolescent Idiopathic Scoliosis: A Case Control Pilot Study

Barbara Dobies-Krześniak [1,*], Agnieszka Werblińska [1] and Beata Tarnacka [2]

[1] Rehabilitation Clinic, 1st Faculty of Medicine, Medical University of Warsaw, Spartańska 1, 02-637 Warsaw, Poland; klinreh@wum.edu.pl

[2] Rehabilitation Clinic, National Institute of Geriatrics, Rheumatology and Rehabilitation, Spartańska 1, 02-637 Warsaw, Poland; klinika.rehabilitacji@spartanska.pl

* Correspondence: bdobies@wum.edu.pl; Tel.: +48-22-6709481

Abstract: The aim of this study was to assess the hypothesis that functional laterality features are associated with scoliosis incidence. The study included 59 patients with radiologically confirmed idiopathic scoliosis (mean age 13 years, 41 girls and 18 boys) and 55 controls (mean age 10.5 years, 38 girls and 17 boys). Side dominance was determined by the Lateral Preference Inventory. Direction, strength, and consistency of lateral dominance was obtained. Continuous data were compared by Student's t-test or U Mann-Whitney test where appropriate. Categorical data were compared by chi-squared test and Fisher's exact test. Groups were significantly different in terms of age ($p < 0.001$) and dependent variables: height ($p < 0.001$) and weight ($p < 0.001$). Lateralization analysis showed some trends, but the results obtained were not statistically significant. Statistical significance of lateralization direction are respectively: for hand ($p = 0.364$); leg ($p = 0.277$); eye ($p = 0.804$); ear ($p = 0.938$); number of right/left sided participants $p = 0.492$; $p = 0.274$; $p = 0.387$; $p = 0.839$, and right/mixed/left sided participants $p = 0.930$; $p = 0.233$; $p = 0.691$; $p = 0.804$. For laterality consistency depending on definition used, $p = 0.105$; $p = 0.108$; $p = 0.380$. The relationship between scoliosis and laterality is not a simple causal relationship and needs further investigation.

Keywords: handedness; sidedness; brain asymmetry; children posture; side dominance

Citation: Dobies-Krześniak, B.; Werblińska, A.; Tarnacka, B. Lateralization Direction, Strength, and Consistency in Juvenile and Adolescent Idiopathic Scoliosis: A Case Control Pilot Study. *Symmetry* **2022**, *14*, 888. https://doi.org/10.3390/sym14050888

Academic Editor: Thierry Paillard

Received: 1 March 2022
Accepted: 20 April 2022
Published: 26 April 2022

1. Introduction

The human body is built on the principle of lateral symmetry. There are exceptions to this principle both in the structure of the body and in the functions of individual organs. Typically, there is clear distribution of functions between sides. Laterality or side dominance is described as a clear advantage of one side of the body over the other in terms of usability, precision, and coordination [1]. Laterality is a characteristic that develops gradually with age and general motor development. The final sensory and motor side dominance is determined around the age of 7 [1,2]. A reliable and valid assessment of laterality is important. There are many definitions of lateralization, and often even within a single definition we may encounter several interpretations [3].

The current literature mainly uses three types of lateralization assessment tools: performance tasks, preference tasks, and self-report questionnaires. Performance tasks compare the quality of tasks performed with both left and right sides. Preference tasks are elicitation of motor responses as an indicator of laterality. Self-report questionnaires gather information about preferences in various motor activities [3]. Due to the heterogeneity of approaches, no standardized examples of "best practices" for assessing laterality dominance are available, nor is there a single definition of "laterality" [3]. Self-report questionnaires are the easiest and most commonly used form of lateralization testing [3].

Asymmetrical spinal load associated with lateral preference is often highlighted as passible contributing factor to scoliosis pathogenesis [4]. Scoliosis is a tri-planar deformity of the spine with Cobb angle lateral curvature of at least 10 degrees (according to Cobb angle), rotation and deformation of the vertebrae. When no cause for the defect can be identified, a diagnosis of idiopathic scoliosis (IS) is made. The baseline prevalence of IS is 2 to 3%. It is more common in girls (♀:♂1.4:1 for angles 10–20° to 7.2:1 for angles >30°) [4]. The etiopathogenesis of IS is still a topic of exploration. Familial incidence suggests a genetic etiology. Among possible causes are abnormalities in estrogen receptor structure and function, mucopolysaccharide, lipoprotein, melatonin or calmodulin synthesis, matrix metalloproteinase-3 (MMP-3) and interleukin-6 (IL-6) promoter polymorphisms, and increased expression of the basonuclin 2 (BNC2) gene [4].

Some prior investigations indicate a significant correlation between direction of hand preference, strength of the asymmetry direction or side preference consistency and incidence of trunk asymmetry, scoliosis, or curve pattern of scoliosis convexity in children and adolescents [5–10]. Others did not confirm these observations [11]. Furthermore, there is no literature on the association between crossed laterality and scoliosis. Crossed laterality can be identified in people who have dominant organs located on opposite sides of the body [1]. This property requires intense cooperation between the hemispheres and can contribute to functional imbalance [1].

This study was designed to try to answer the question: can a significantly different level of lateralization traits (direction, strength and consistency) be identified in children with scoliosis?

2. Materials and Methods

2.1. Study and Control Group Background

The study group was recruited among patients of the Orthopedic and Rehabilitation Centre for Children and Youth admitted between January 2018 and July 2019. The study included children and adolescents aged 7–18 years. Patients with IS with Cobb angle ≥10° and vertebral rotation on anteroposterior radiograph taken in the last 6 months without connective tissue disorder in medical history and their legal guardians were approached. Informed written consent was obtained.

The control group aged 7–15 was recruited from volunteers from St. Francis School in Warsaw, Poland and was examined from April to June 2019. Children with abnormal posture diagnosed prior to study inclusion (abnormalities in spinal shape with suggestion of therapeutic intervention) were excluded. The study protocol was approved by the Warsaw Medical University Ethic Committee and was conducted in accordance with the Declaration of Helsinki principles. Study and control groups were matched by sex ratio [8,9].

The demographic and medical data (including medical history, age, sex, weight, hight) was collected from both groups.

No a priori sample estimation was performed.

2.2. Scoliosis Angular Value

To assess scoliosis in the study group, lateral spinal curve size was measured using the Cobb method [12]. BDK, using the Radiant Dicom Viewer computer program, drew lines parallel to the upper border of the upper vertebral body and the lower border of the body of the lowest vertebra of the structural arch (the vertebra most deviated from vertical), obtaining the angular value of the scoliosis.

2.3. Lateral Preference Inventory (LPI)

Side dominance was determined by the Lateral Preference Inventory (LPI) [13]. The Polish version of LPI was translated by the first author and approved by the other authors. The survey, which consists of items to assess hand, foot, eye, and ear preference was filled by the subjects. Each item requires the response of "left", "right", or "either" [13].

2.3.1. Side Dominance Direction and Strength

To analyze the data from the study, we scored 1 for each "right", "0" for "either", and "−1" for "left" answer, giving a score for each four-item scale from −4 to 4, with

−4 meaning consistent left-sidedness and 4 meaning consistent right-sidedness for any index. Then, laterality data were categorized as a dichotomy (R/L; right/left), where R is the number of "right" responses (1 to 4) and L is the number of "left" (0 to −4), and a trichotomy (right/ mixed/left; R/M/L) where "R" describes consistent right laterality with score of 4, "M"—mixed or weak laterality—from 3 to −3, and "L"—consistent left with score of −4.

2.3.2. Side Dominance Consistency

To compare the prevalence of crossed laterality, study participants were grouped based on using the opposite sides of the body while performing different tasks with any combination of hand, eye, foot, or ear [14]. We considered three used definitions of crossed laterality. Consistent (absolute) crossed laterality with at least one score of consistent right (=4), and at least one consistent left (score of −4). Secondly, we compared prevalence of strong crossed laterality with at least one result of strong right (3/4 tests marked "1") and at least one strong left (3/4 tests marked "−1") in both groups. Thirdly, simple (relative) crossed laterality with at least one "R" (scored from 4 to 1) and one "L" (from 0 to −4) item was compared between groups.

2.4. Statistics

For statistical analysis, mean value and standard deviation (SD) were used to present continuous data. Categorical data were presented as a percentage. The Kolmogorov-Smirnov test was used to verify normal distribution. Continuous data were compared by Student's *t*-test or U Mann-Whitney test where appropriate. Categorical data were compared by chi-squared test and Fisher's exact test. For statistical significance, *p* value less than 0.05 was considered. The statistician used IBM SPSS Statistics for Windows, v21 (IBM Corp., Armonk, NY, USA)

G*Power software v.3.1.9.4 was used to determine the effect size and conduct a post hoc analysis of the power of the study.

3. Results

3.1. Basic Statistics of the Studied Groups

The study group consisted of 59 patients with radiologically confirmed IS (mean age 13 years, 41 girls and 18 boys). A control group with 55 subjects (mean age 10.5 years, 38 girls and 17 boys) participated in this study. Basic parameters comparing both groups are shown in Table 1.

Table 1. Basic parameters of the study and the control groups.

	Study Group (n = 59)	Control Group (n = 55)	p (<0.05)
Age (years)	13.0 ± 2.4	10.5 ± 2.1	<0.001
Male sex (%)	18 (30.5)	17 (30.9)	0.963
Female sex (%)	41 (69.5)	38 (69.1)	0.972
Weight (kg)	52.3 ± 16.2	35.5 ± 10.4	<0.001
Height (cm)	160.3 ± 13.9	144.3 ± 13.3	<0.001

3.2. Comparative Analysis of Direction of Laterality between the Study Group and the Control Group

Table 2 shows the comparison between the study group and the control group in terms of laterality for each item and for the summed mean total numerical score (from −4 to 4) for each four-item scale. In general, we can see a tendency to laterally shift to the left in the study group for handedness and footedness. This effect did not reach the level of statistical significance in our study. The only item with statistical significance (identifying the ear with which the child would listen to the heartbeat) indicates a less pronounced dominance of the left ear in the scoliosis group.

Table 2. Comparative analysis of LPI items and mean for each scale.

Lateral Preference Inventory	Study Group n = 59 (100%)			Control Group n = 55 (100%)			p (<0.05)	Effect Size
	1	0	−1	1	0	−1		
Drawing	54 (91.5)	1 (1.7)	4 (6.8)	51 (92.7)	3 (5.5)	1 (1.8)	0.253	
Hit a target	40 (67.8)	15 (25.4)	4 (6.8)	43 (78.2)	9 (16.4)	3 (5.5)	0.446	
Eraser	47 (79.7)	6 (10.2)	6 (10.2)	49 (89.1)	2 (3.6)	4 (7.3)	0.316	
Dealing cards	46 (78.0)	6 (10.2)	7(11.9)	41 (74.5)	8 (14.5)	6 (10.9)	0.775	
Handedness		2.8 ± 1.8			3.1 ± 1.4		0.364	d = 0.186
Kicking a ball	46 (78.0)	10 (16.9)	3 (5.1)	42 (76.4)	8 (14.5)	5 (9.1)	0.682	
Pick up a pebble	29 (49.2)	27 (45.8)	3 (5.1)	37 (67.3)	15 (27.3)	3 (5.5)	0.119	
Stepping on a bug	33 (55.9)	22 (37.3)	4 (6.8)	36 (65.5)	14 (25.5)	5 (9.1)	0.390	
Stepping up	45 (76.3)	8 (13.6)	6(10.2)	49 (89.1)	1 (1.8)	5 (9.1)	0.062	
Footedness		2.3 ± 1.6			2.7 ± 1.7		0.277	d = 0.242
Telescope	39 (66.1)	9 (15.3)	11 (18.6)	34 (61.8)	6 (10.9)	15 (27.3)	0.492	
Looking into	38 (64.4)	8 (13.6)	13 (22.0)	41 (75.5)	4 (7.3)	10 (18.2)	0.427	
Keyhole	44 (74.6)	3 (5.1)	12 (20.3)	38(69.1)	1 (1.8)	16 (29.1)	0.392	
Sighting a rifle	36 (61.0)	14 (23.7)	9 (15.3)	37 (67.3)	6 (10.9)	12 (21.8)	0.173	
Eyedness		1.9 ± 2.7			1.8 ± 3.1		0.804	d = 0.034
Eavesdropping	33 (55.9)	17 (28.8)	9 (15.3)	35 (63.6)	12 (21.8)	8 (14.5)	0.657	
Earphone	37 (62.7)	17 (28.8)	5 (8.5)	34 (61.8)	15 (27.3)	6 (10.9)	0.904	
Heartbeat	34 (57.6)	19 (32.2)	6 (10.2)	29 (52.7)	11 (20.0)	15 (27.3)	0.044	
Clock in the box	37 (62.7)	14 (23.7)	8 (13.6)	42 (76.4)	9 (16.4)	4 (7.3)	0.273	
Earedness		1.9 ± 1.9			1.9 ± 2.2		0.938	d = 0.000

In dichotomized groups R/L (Table 3), hand and foot dominance items tested showed a trend to shift to the left in the study group, and eye and ear tended to shift to the right. Again, these effects did not achieve statistical significance.

Table 3. Comparative analysis of direction of laterality.

Dominance	Study Group n = 59 (100%)		Control Group n = 55 (100%)		p (<0.05)
	R	L	R	L	
Hand	53 (89.8)	6 (10.2)	52 (94.5)	3 (5.5)	0.492
Leg	51 (86.4)	8 (13.6)	51 (92.7)	4 (7.3)	0.274
Eye	46 (78.0)	13 (22.0)	39 (70.9)	16 (29.1)	0.387
Ear	46 (78.0)	13 (22.0)	42 (76.4)	13 (23.6)	0.839

3.3. Comparative Analysis of Strength of Laterality between the Study Group and the Control Group

The influence of laterality strength on the occurrence of scoliosis was assessed. The trichotomy R/M/L considering patients with pure right or left laterality for hand, leg, eye, or ear was analyzed. For all four subdomains, more subjects with weak laterality could be found in the scoliosis group (Table 4), but the effect observed is not statistically significant.

Table 4. Comparative analysis of strength of laterality.

Dominance:	Study Group n = 59 (100%)			Control Group n = 55 (100%)			p (<0.05)
	R	M	L	R	M	L	
Hand	28 (47.5)	30 (50.8)	1 (1.7)	28 (50.9)	26 (47.3)	1 (1.8)	0.930
Leg	15 (25.4)	43 (72.9)	1 (1.7)	21 (38.2)	34 (61.8)	0 (0.0)	0.233
Eye	27 (45.8)	26 (44.1)	6 (10.2)	28 (50.9)	20 (36.4)	7 (12.7)	0.691
Ear	15 (25.4)	43 (72.9)	1 (1.7)	17 (30.9)	37 (67.3)	1 (1.8)	0.804

3.4. *Comparative Analysis of Crossed Laterality between the Study Group and the Control Group*

When analyzing the consistency in preference, we can see a general trend towards more frequent crossed laterality prevalence in the control group. This observation does not have statistical significance (Table 5).

Table 5. Comparative analysis of crossed laterality.

Crossed Laterality	Study Group n = 59 (100%)	Control Group n = 55 (100%)	p (<0.05)
Consistent	1 (1.7)	5 (9.1)	0.105
Strong	8 (13.6)	14 (25.5)	0.108
Simple	21 (35.6)	24 (43.6)	0.380

4. Discussion

To our knowledge, our study is the only one that examines the prevalence and characteristics of not only hand and leg, but also eye and ear laterality using multi-item inventory in radiologically confirmed scoliotic patients with a control group.

The prevalence of left-handedness is reported between 1% and 30% depending on age, sex, handedness testing method, nationality, and sociological characteristics [3]. In our study, the frequency of left-handedness was 10.2% for the study group and 5.5% for the control group (Table 3), but the difference was not statistically significant (p = 0.492). The effect sizes of summed mean total numerical score for hand and leg laterality were d = 0.186 and d = 0.242 (small effect) and for eye and ear d = 0.034 and d = 0.000 (neglectable effect). Based on the values obtained, a post hoc analysis of the power of the study was performed, giving values of ~0.75 for hand and leg and >0.85 for eye and ear.

The main issue may not be the direction of preference alone, but the strength of laterality. Dominance appears weaker lateralized among younger respondents [15–17]. Children with clear hand dominance show less coordination problems than their poorly lateralized peers [18]. Also, early development of strong dominance (regardless of dominating side) correlate with better coordination [19]. In our study, children with pure laterality for all items tested appeared more frequently in the control group, despite their younger age (Table 4). Again, this observation was not supported by statistical significance and is opposite to observation from the study of Goldberg and Dowling [5].

Crossed laterality is a topic that needs a closer look. Inconsistencies in any pair of lateral preferences can be noted in 69.2% of healthy adults [15]. Consistency in preference across different domains increases among older respondents [15–17].

In our study, more subjects with crossed laterality of any dominance pair can be found in the control group, irrespective of the definition of crossed laterality adopted. This tendency did not achieve statistical significance (Table 5).

Results should be considered, keeping in mind the age difference in our study. The study group would be even more left lateralized and even less strongly lateralized when considering a younger population. In turn, the lateralization intersection difference would likely lose value when the groups were equalized by age.

There is no consensus in the literature whether the occurrence of scoliosis is directly related to the side and strength of lateral dominance [5–11].

Grivas et al. examined 8245 children 6 to 18 years old. Significant correlation between handedness and trunk asymmetry in the group of 2–7° mid-thoracic asymmetry was noted [10]. Chiara et al. examined 1029 Italian children aged 11–14 years. The left-side dominance was marked as a possible predictor of trunk asymmetry in thoracic and thoracolumbar curves [9].

Milenkovic et al. investigated a group of 2546 children 11 to 14 years old. Co-occurrence of left-handedness and scoliosis was statistically significant in girls [8]. Goldberg and Dowling studied 254 girls for scoliosis convexity association with handedness. The correlation of curve pattern and handedness was noted in 82% of the cases and was statistically significant.

However, in comparison to normal population proportion of left-handers among IS patients was typical [6]. The two authors have continued the topic in 1991, examining 159 children with IS using the questionnaire by Porac and Coren for hand and leg preference. They concluded that scoliotic patients tend to be more strongly lateralized than healthy peers [5]. In 2006, they published large study of 1636 children, of whom 673 had IS. The work showed a significant correlation between hand preference and scoliosis pattern [7].

All the papers mentioned differ from ours in terms of methodology. Some identify scoliosis only in terms of an abnormal Adams test result [8–11] or lateral curvature in radiogram as low as 5 degrees [6]. Such planning of the studies was probably aimed at obtaining a larger study group without the burden of the radiological examination.

There is a possibility, that lefthanders trying to adapt to a right-handed world during their daily activities at home, school, or in their social environment adopt incorrect postures developing trunk asymmetries with abnormal Adams test, but this effect does not lead to changes in bone structure of the spine [6].

Identification of dominant side is also inconsistent among studies. It could be caused by multiplicity of diagnostic methods and the heterogeneity of nomenclature.

In most previous works, only dominant (or writing) hand was identified. The identification was typically based on a question to the child or parent of which hand is preferred [6–10]. In only two papers the authors used a survey questionnaire to determine lateralization. [5,11]. This simplification allowed for larger study groups to be recruited.

In the recommendations, we can find a suggestion that multi-item inventory should be used [15]. Also, response categories consisting of 'right', 'left', and 'no preference' are considered sufficiently accurate for the assessment of lateral preferences [15]. LPI was the only survey indicated to be reliable [15].

Limitations

The most important limitation of the study is the age difference between the two groups because the mixed-siders are generally younger than both right- and left-siders [15], but this trend is most relevant when considering children under 7 years of age [16].

The age difference between the groups also explains the statistically significant differences in weight and height.

For this topic, sample size and survey power are challenging due to the high asymmetry of the lateral preference itself and the skewed distributions of consistency and strength of lateral preference. Cobb angle measurement for the study group was performed by one researcher (BDK) in a single measurement. The intra-rater reliability for this investigator was 0.706 when assuming a $\pm 1^{\circ}$ range and 0.941 when assuming a $\pm 2^{\circ}$ range.

There is a methodological problem across the studies—a lack of agreement for the definition, instruments, and methods to assess lateral preference.

5. Conclusions

The relationship between radiologically confirmed scoliosis and laterality is inconclusive.

Our study, due to the lack of statistical significance of the observations made, does not provide clinically relevant conclusions. However, it shows trends that require further observation in larger study groups, from our calculations for 0.8 power of the study and $p = 0.05$ for hand and leg lateralization minimum 359 participants in each group, keeping in mind the age difference in our study.

Author Contributions: Conceptualization, B.D.-K.; data curation, B.D.-K. and A.W.; formal analysis, B.D.-K.; funding acquisition, B.D.-K.; investigation, B.D.-K. and A.W.; methodology, B.D.-K. and B.T.; project administration, B.D.-K.; resources, B.D.-K.; supervision, B.D.-K. and B.T.; visualization, B.D.-K.; writing—original draft, B.D.-K.; writing—review and editing, B.D.-K., A.W. and B.T. All authors have read and agreed to the published version of the manuscript.

Funding: This research received no external funding.

Institutional Review Board Statement: The study was conducted according to the guidelines of the Declaration of Helsinki and acknowledged by the Ethics Committee of Warsaw Medical University (KB/38/2017 from 7 March 2017).

Informed Consent Statement: The informed written consent to participate in the study was obtained from all subjects and their legal guardians. All data is anonymized. Individual consent for publication is not required.

Data Availability Statement: The datasets used and/or analyzed during the current study are available from the corresponding author on reasonable request.

Acknowledgments: Piotr Tederko, Justyna Frasuńska.

Conflicts of Interest: The authors declare no conflict of interest.

Abbreviations

IS—idiopathic scoliosis.

References

1. Osiński, W. *Antropomotoryka*, 2nd ed.; AWF Poznań: Poznań, Poland, 2003; pp. 286–295.
2. Bondi, D.; Prete, G.; Malatesta, G.; Robazza, C. Laterality in Children: Evidence for Task-Dependent Lateralization of Motor Functions. *Int. J. Environ. Res. Public Health* **2020**, *17*, 6705. [CrossRef] [PubMed]
3. Utesch, T.; Mentzel, S.; Strauss, B.; Büsch, D. Chapter 4—Measurement of laterality and its relevance for sports. In *Laterality in Sports. Theories and Applications*; Loffing, F., Hagemann, N., Strauss, B., MacMahon, C., Eds.; Elsevier: Amsterdam, The Netherlands; Academic Press: Cambridge, MA, USA, 2016; pp. 65–86. [CrossRef]
4. Negrini, S.; Donzelli, S.; Aulisa, A.G.; Czaprowski, D.; Schreiber, S.; de Mauroy, J.C.; Diers, H.; Grivas, T.B.; Knott, P.; Kotwicki, T.; et al. 2016 SOSORT guidelines: Orthopaedic and rehabilitation treatment of idiopathic scoliosis during growth. *Scoliosis Spinal Disord.* **2018**, *13*, 3. [CrossRef] [PubMed]
5. Goldberg, C.J.; Dowling, F.E. Idiopathic scoliosis and asymmetry of form and function. *Spine* **1991**, *16*, 84–87. [CrossRef] [PubMed]
6. Goldberg, C.; Dowling, F.E. Handedness and scoliosis convexity: A reappraisal. *Spine* **1990**, *15*, 61–64. [CrossRef] [PubMed]
7. Goldberg, C.J.; Moore, D.P.; Fogarty, E.E.; Dowling, F.E. Handedness and spinal deformity. *Stud. Health Technol. Inform.* **2006**, *123*, 442–448. [PubMed]
8. Milenkovic, S.; Kocijancic, R.; Belojevic, G. Left handedness and spine deformities in early adolescence. *Eur. J. Epidemiol.* **2004**, *19*, 969–972. [CrossRef] [PubMed]
9. Arienti, C.; Buraschi, R.; Donzelli, S.; Zaina, F.; Pollet, J.; Negrini, S. Trunk asymmetry is associated with dominance preference: Results from a cross-sectional study of 1029 children. *Braz. J. Phys. Ther.* **2019**, *23*, 324–328. [CrossRef] [PubMed]
10. Grivas, T.B.; Vasiliadis, E.S.; Polyzois, V.D.; Mouzakis, V. Trunk asymmetry and handedness in 8245 school children. *Pediatr. Rehabil.* **2006**, *9*, 259–266. [CrossRef] [PubMed]
11. Fernandez, M.; Fernandez, R.; Zurita, F.; Jimenez, C.; Almagia, A.; Yuing, T.; Curilem, C. Relationship between Scoliosis, Sex and Handedness in a Sample of Schoolchildren. *Int. J. Morphol.* **2015**, *33*, 24–30.
12. Cobb, J.R. The problem of the primary curve. *J. Bone Joint Surg. Am.* **1960**, *42*, 1413–1425. [CrossRef] [PubMed]
13. Coren, S. The Lateral Preference Inventory for measurement of handedness, footedness, eyedness, and earedness: Norms for young adults. *Bull. Psychon. Soc.* **1993**, *31*, 1–3. [CrossRef]
14. Ferrero, M.; West, G.; Vadillo, M.A. Is crossed laterality associated with academic achievement and intelligence? A systematic review and meta-analysis. *PLoS ONE* **2017**, *12*, e0183618. [CrossRef]
15. Tran, U.S.; Stieger, S.; Voracek, M. Evidence for general right-, mixed-, and left-sidedness in self-reported handedness, footedness, eyedness, and earedness, and a primacy of footedness in a large-sample latent variable analysis. *Neuropsychologia* **2014**, *62*, 220–232. [CrossRef] [PubMed]
16. Greenwood, J.G.; Greenwood, J.J.; McCullagh, J.F.; Beggs, J.; Murphy, C.A. A survey of sidedness in Northern Irish schoolchildren: The interaction of sex, age, and task. *Laterality* **2007**, *12*, 1–18. [CrossRef] [PubMed]
17. Suar, D.; Mandal, M.K.; Misra, I.; Suman, S. Lifespan trends of side bias in India. *Laterality* **2007**, *12*, 302–320. [CrossRef] [PubMed]
18. Hill, E.L.; Bishop, D.V. A reaching test reveals weak hand preference in specific language impairment and developmental co-ordination disorder. *Laterality* **1998**, *3*, 295–310. [CrossRef] [PubMed]
19. Gabbard, C.; Hart, S.; Kanipe, D. Hand preference consistency and fine motor performance in young children. *Cortex* **1993**, *29*, 749–753. [CrossRef]

symmetry

MDPI

Article

Kinematic Analysis of Lower Limb Joint Asymmetry During Gait in People with Multiple Sclerosis

Massimiliano Pau [1,*], Bruno Leban [1], Michela Deidda [1], Federica Putzolu [1], Micaela Porta [1], Giancarlo Coghe [2] and Eleonora Cocco [2]

1 Department of Mechanical, Chemical and Materials Engineering, University of Cagliari, 09124 Cagliari, Italy; bruno.leban@dimcm.unica.it (B.L.); m.deidda32@tiscali.it (M.D.); federica.putzolu98@tiscali.it (F.P.); m.porta@dimcm.unica.it (M.P.)

2 Multiple Sclerosis Centre, Department of Medical Sciences and Public Health, University of Cagliari, 09124 Cagliari, Italy; gccoghe@gmail.com (G.C.); ecocco@unica.it (E.C.)

* Correspondence: massimiliano.pau@dimcm.unica.it; Tel.: +39-070-675-3264

Abstract: The majority of people with Multiple Sclerosis (pwMS), report lower limb motor dysfunctions, which may relevantly affect postural control, gait and a wide range of activities of daily living. While it is quite common to observe a different impact of the disease on the two limbs (i.e., one of them is more affected), less clear are the effects of such asymmetry on gait performance. The present retrospective cross-sectional study aimed to characterize the magnitude of interlimb asymmetry in pwMS, particularly as regards the joint kinematics, using parameters derived from angle-angle diagrams. To this end, we analyzed gait patterns of 101 pwMS (55 women, 46 men, mean age 46.3, average Expanded Disability Status Scale (EDSS) score 3.5, range 1–6.5) and 81 unaffected individuals age- and sex-matched who underwent 3D computerized gait analysis carried out using an eight-camera motion capture system. Spatio-temporal parameters and kinematics in the sagittal plane at hip, knee and ankle joints were considered for the analysis. The angular trends of left and right sides were processed to build synchronized angle–angle diagrams (cyclograms) for each joint, and symmetry was assessed by computing several geometrical features such as area, orientation and Trend Symmetry. Based on cyclogram orientation and Trend Symmetry, the results show that pwMS exhibit significantly greater asymmetry in all three joints with respect to unaffected individuals. In particular, orientation values were as follows: 5.1 of pwMS vs. 1.6 of unaffected individuals at hip joint, 7.0 vs. 1.5 at knee and 6.4 vs. 3.0 at ankle ($p < 0.001$ in all cases), while for Trend Symmetry we obtained at hip 1.7 of pwMS vs. 0.3 of unaffected individuals, 4.2 vs. 0.5 at knee and 8.5 vs. 1.5 at ankle ($p < 0.001$ in all cases). Moreover, the same parameters were sensitive enough to discriminate individuals of different disability levels. With few exceptions, all the calculated symmetry parameters were found significantly correlated with the main spatio-temporal parameters of gait and the EDSS score. In particular, large correlations were detected between Trend Symmetry and gait speed (with rho values in the range of −0.58 to −0.63 depending on the considered joint, $p < 0.001$) and between Trend Symmetry and EDSS score (rho = 0.62 to 0.69, $p < 0.001$). Such results suggest not only that MS is associated with significantly marked interlimb asymmetry during gait but also that such asymmetry worsens as the disease progresses and that it has a relevant impact on gait performances.

Keywords: gait; kinematics; spatio-temporal; multiple sclerosis (MS); cyclograms; angle-angle diagrams; symmetry

Citation: Pau, M.; Leban, B.; Deidda, M.; Putzolu, F.; Porta, M.; Coghe, G.; Cocco, E. Kinematic Analysis of Lower Limb Joint Asymmetry During Gait in People with Multiple Sclerosis. *Symmetry* **2021**, *13*, 598. https://doi.org/10.3390/sym13040598

Academic Editor: Chiarella Sforza

Received: 10 March 2021
Accepted: 1 April 2021
Published: 3 April 2021

1. Introduction

Multiple Sclerosis (MS) is a chronic immunomediated and neurodegenerative disease of the central nervous system, which represents the most frequent cause of disability among young adults [1–3]. Being characterized by symptoms such as weakness, fatigue and spasticity, MS can significantly compromise the efficient performance of several basic motor functions, including postural control [4], locomotion [5], upper extremity capabilities [6] and, in general, several common activities of daily living (ADL [7]). In people with MS (pwMS), lower limb motor dysfunctions, although present in both limbs, are usually asymmetrical in magnitude. This is especially true regarding self-perceived weakness, muscular strength and activity, power and limb loading [8,9], but spasticity is also characterized by unilateral presentation in a non-negligible percentage (estimated between 10 and 16%) of pwMS [10].

Since pwMS primarily complain about weakness, a relevant number of studies have attempted to objectively quantify the existence of actual interlimb muscle strength asymmetries. These were indeed almost unanimously found, especially at knee level [11–15]. However, it remains unclear how, and to what extent, they impact motor tasks that rely on optimal bilateral coordination such as balance and gait. As pointed out in the recent review by Rudroff and Proessl [9], although some studies report significant associations between muscle function asymmetries, postural stability and walking performance, others do not. It has been suggested that such inconsistencies are due to wide variability in asymmetry assessment methods [9] but it is also to be considered that in many cases, asymmetry assessment is separately performed with respect to the specific motor task that is supposedly affected by it. Thus, the role of muscle function cannot be analyzed in a true ecological context.

In contrast, less explored appears the effect of the disease in terms of joint kinematics asymmetry as, to the best of our knowledge, only a few studies have explicitly investigated the existence of possible differences in mobility of lower limb joints. Daunoraviciene et al. [16] employed inertial sensors to assess asymmetry of lower limb joints in pwMS who carried out the heel-to-shin test, while Filli et al. [17] analyzed the existence of interlimb differences in the range of motion (ROM) at hip, knee and ankle joints during a gait analysis using an optical motion capture system in a study aimed to profile walking dysfunctions on pwMS. Crenshaw et al. [18] employed the angular trend waveforms in the sagittal plane for hip, knee, and ankle joints during gait to determine several gait symmetry measures (i.e., trend similarity, phase shift, minimum trend similarity, range amplitude ratio, and range offset) using an eigenvector approach. They reported that pwMS were generally more asymmetrical than unaffected individuals and that asymmetry parameters worsened in the fatigued condition. Since knowledge of lower limb kinematics has been recognized as an essential factor in better understanding the underlying mechanisms of walking disability in MS, [18], it is reasonable to hypothesize that the availability of data on joint movement asymmetry would be extremely pertinent to quantify the magnitude of its impact on walking performance.

1.1. Characterization of Gait Asymmetry in PwMS: Methods Based on Discrete Values

The study of gait asymmetry in pwMS is usually performed through analysis of differences between more affected and less affected limbs in terms of spatio-temporal parameters. To this end, symmetry is quantified by means of several parameters, among which the most used is represented by the Symmetry Index (SI, originally proposed by Robinson et al. [19]), which is expressed by the following equation:

$$SI = \frac{2 \times (V_{la} - V_{ma})}{(V_{la} + V_{ma})} \times 100 \tag{1}$$

where V_{la} and V_{ma} represent the values of the gait variable of interest (usually step time, step length, or duration of stance, swing and single and double support phases), calculated, respectively, for the less affected and the more affected limb or, in a more general formulation, for the left and right side. When no differences are measured between the two limbs, SI becomes null and gait is considered perfectly symmetric, while as SI increases, asymmetry increases. The original formulation by Robinson has been subsequently modified by other authors (see the review by Viteckova et al. [20] for details) to adapt it to different gait variables. Values of SI during gait for pwMS have been reported as regards studies on the characterization of gait pattern for different MS phenotypes [21] and as outcome of rehabilitative treatments [22].

It is noteworthy that other sophisticated approaches, such as nonlinear ones based on either multiresolution entropy [23] or cross-fuzzy entropy [24], have been proposed to investigate lower limb symmetry in individuals affected by neurologic conditions. Like discrete methods, nonlinear methods are based on the calculation of discrete variables extracted from a continuous signal to perform the assessment of symmetry, through evaluation of the evolution of a discrete variable over a set of consecutive gait cycles [20].

1.2. Waveform-Based Methods to Assess Interlimb Symmetry During Gait

Since discrete approaches previously described focus on a single or a limited set of events, they are unable to provide information on the way a certain kinematic variable (and thus asymmetry) evolves over time. This drawback can be overcome by using waveform-based methods that exploit all kinematic information contained in the curve of variation of the lower limb joint angles with time during a complete gait cycle. However, this increase in information content and accuracy comes at a cost: waveform-based techniques are more complex to implement and time-consuming. Moreover, interpretation of the parameters they provide is not so straightforward as occurs with classic symmetry indexes. However, several studies carried out in the last decade on individuals affected by neurologic (neuropathies, stroke, Parkinson's disease [25–27]) and orthopedic conditions [28–30] demonstrated that such an approach is versatile and allows a more accurate and thorough analysis of gait symmetry, thus proving to be of great relevance in all conditions characterized by subtle, not easily detectable alteration of gait with the conventional discrete indices.

One of the best-known and most widespread methods for investigating symmetry either between the same joint of left and right lower limb or between two joints of the same limb is based on analysis of angle–angle diagrams, also known as "cyclograms". Originally proposed by Grieve in 1968 [31], they rapidly attracted the interest of researchers and clinicians, since symmetry was graphically and mathematically expressed through simple geometrical properties of the figures generated by the angle–angle comparison such as area, perimeter, etc. In the last two decades, more refined mathematical approaches have been formulated to make the method sensitive to even relatively low asymmetries.

Surprisingly, although the impact of asymmetry issues associated with MS is extremely relevant, the literature reports only one study (carried out on a small sample of 13 pwMS) in which angle–angle diagrams and associated summary parameters were used [32]. Its major findings were a more marked asymmetry of pwMS with respect to unaffected individuals and the absence of significant relationships between the level of disability and the symmetry parameters. Considering the informative potential of this approach and the substantial lack of data, we propose here a retrospective study performed on a large cohort of pwMS who underwent a computerized 3D gait analysis during a 5-year period. In particular, the main purposes of the research are as follows: (1) to employ waveform-based methods to assess lower-limb joint kinematics asymmetry during gait in a cohort of pwMS and verify whether the values of the calculated symmetry parameters are significantly different from those of unaffected individuals or not; (2) to assess the existence of possible differences in asymmetry between pwMS characterized by different levels of disability; and (3) to verify the existence of possible relationships between asymmetry parameters and spatio-temporal parameters of gait and disability level. A secondary goal of the study is to compare the ability of different indicators associated with angle–angle diagrams to correctly discriminate pwMS from unaffected individuals and pwMS between them depending on their level of disability.

2. Materials and Methods

2.1. Participants

In the period May 2014 February 2020, 236 pwMS followed at the Regional Multiple Sclerosis Center of Sardinia (Cagliari, Italy) underwent a computerized three-dimensional gait analysis at the Laboratory of Biomechanics and Industrial Ergonomics of the University of Cagliari (Cagliari, Italy). They had previously been diagnosed with MS by a neurologist expert in MS (E.C., G.C.) according to the 2010 revised criteria [33,34] and tested in the laboratory to either characterize and monitor alterations of gait associated with the disease progression or assess the effect of pharmacologic and rehabilitative treatments [35–37]. For the purposes of the present study, only pwMS able to ambulate autonomously (i.e., without the support of canes, crutches or walking frames) for at least 100 m and free from any other condition potentially able to severely affect gait or balance were considered. Such a selection, which resulted in a sub-group composed of 101 unique pwMS (55 women, 46 men, mean age 46.3 years) was carried out to remove any possible confounding effects on gait kinematics associated with the presence of walking aids [38,39]. In the case of pwMS who were recruited for interventional studies, the test condition considered for the present analysis was the "pre-intervention".

Participants were stratified into two groups depending on their disability level assessed through the Expanded Disability Status Scale (EDSS) score as follows:

- Low-mild disability (EDSS ≤ 3.5, $n = 59$)
- Moderate-severe disability (EDSS > 3.5, $n = 42$)

Eighty-one unaffected individuals age- and sex-matched recruited among nurses and staff of the MS Center and the University of Cagliari served as the control group. The main anthropometric and clinical features of all participants are reported in Table 1.

Table 1. Anthropometric and clinical features of participants. Values are expressed as mean (SD).

	Healthy Controls	All MS	MS Low-Mild Disability (EDSS ≤3.5)	MS Moderate-Severe Disability (EDSS >3.5)
Participants (M, F)	81 (44F, 37M)	101 (55F, 46M)	59 (33F, 26M)	42 (22F, 20M)
Age (years)	48.9 (15.2)	46.3 (10.4)	44.2 (10.3)	49.3 (9.7)
Body Mass (kg)	65.2 (11.4)	64.7 (12.0)	66.1 (12.5)	62.8 (11.1)
Height (cm)	167.2 (9.1)	166.3 (9.3)	166.7 (9.6)	165.8 (9.0)
EDSS Score	–	3.5 (1.7)	2.4 (1.0)	5.2 (1.0)

EDSS: Expanded Disability Status Scale; MS: Multiple Sclerosis.

All data presented here wer obtained within several studies conducted according to the principles expressed in the World Medical Association Declaration of Helsinki and formally approved by the local Ethics Committee (authorization numbers 180/2014, 102/2018 and 198/2019). In all cases, participants signed an informed consent agreeing to participate.

2.2. Spatio-Temporal and Kinematic Data Collection and Processing

An optical motion-capture system (Smart-D, BTS Bioengineering, Italy) composed of 8 infrared cameras set at 120 Hz frequency was employed to acquire the trajectories of 22 spherical retro-reflective passive markers (14 mm diameter) placed on the skin of participants' lower limbs and trunk at specific landmarks according to the protocol described by Davis et al. [40]. After the acquisition of main anthropometric data (i.e., height, weight, anterior superior iliac spine (ASIS) distance, pelvis thickness, knee and ankle width and leg length) and the markers' placement, participants walked at a self-selected speed in the most natural manner possible on a 10 m walkway at least 6 times, interspersed with suitable rest times.

The raw data were first processed with the dedicated Smart Analyzer software (BTS Bioengineering, Italy) to calculate the main spatio-temporal parameters of gait (speed, stride length, cadence, step width, stance, swing and duration of double support phases) and derive the mean value of the angles at hip, knee and ankle joint during the gait cycle calculated on the basis of the six trials. Such curves were then exported as ASCII files for further processing with a custom routine developed under Matlab ® environment (see Appendix A), which calculated interlimb symmetry as described later in detail.

2.3. Gait Symmetry Quantification by Means of Cyclograms

Synchronized bilateral cyclograms were generated according to the procedure described by Goswami [41]. To this end, right and left limb angle values acquired during the gait cycle were used to build angle–angle diagrams for each joint of interest (i.e., hip, knee and ankle). A number of geometrical features of cyclograms were then extracted as follows (see also Figure 1 for a graphical explanation):

- Cyclogram area (degrees^2) represents the area of the closed curve obtained from each angle–angle diagram [42]. Since a perfectly symmetrical gait is achieved when both left and right joints are positioned at the same angle for a certain time of the gait cycle (i.e., all the cyclogram points stand on a 45° line in the diagram and thus the area is null), the interpretation of this parameter is quite straightforward; that is, the smaller the area, the more symmetrical the gait.
- Cyclogram orientation (degrees): this feature is identified by the absolute value of angle ϕ formed by the 45° line, which corresponds to perfect interlimb symmetry and the orientation of the principal axis of inertia, which corresponds to the minimum moment of inertia of the cyclogram [41,43]. The latter was calculated as the direction of the eigenvector of the matrix of inertia of the cyclogram points distribution in the x-y (i.e., left joint angle–right joint angle) reference system. Smaller values of this angle indicate higher interlimb symmetry.
- Trend Symmetry: this dimensionless parameter quantifies the similarity between two waveforms (in our case time-normalized right leg and left leg angular trend across the gait cycles for each joint of interest) using an eigenvector analysis (see [44] for details of the mathematical procedure). In particular, it is obtained by dividing the variability about the eigenvector to the variability along the eigenvector, and it is not influenced by the presence of a shift or by magnitude differences in two waveforms. Even in this case, the interpretation of this parameter is quite simple; a 0 value indicates perfect symmetry, and asymmetry increases as the Trend Symmetry value increases.

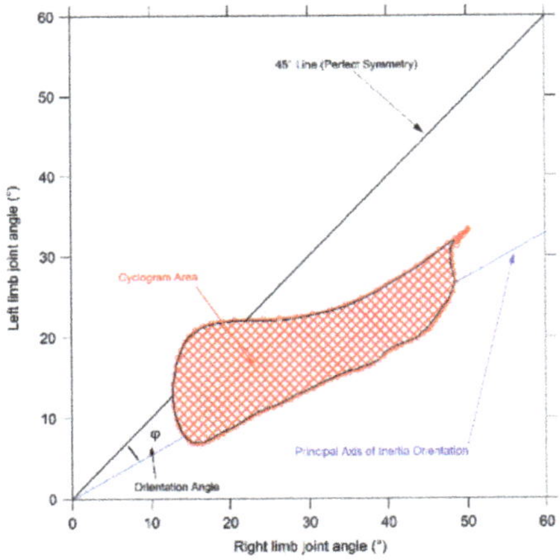

Figure 1. Graphic representation of a cyclogram and its main features considered for the present study.

2.4. Statistical Analysis

Parametric statistical analysis was adopted after preliminarily checking data for normality (using the Shapiro–Wilk's test) and homogeneity of variances (Levene's test). The existence of possible differences in symmetry introduced by the presence of MS was investigated using two distinct one-way multivariate analyses of variance (MANOVA). The first one, which investigated the differences between spatio-temporal parameters of pwMS and unaffected individuals, was performed by considering the participant's status (i.e., unaffected and pwMS with low–mild or moderate–severe disability) as independent variables and the 7 spatio-temporal parameters previously mentioned (speed, stride length, cadence, step width, stance, swing and duration of double support phases) as dependent variables. In the second MANOVA, we analyzed the effect of the presence of MS on symmetry parameters. In this case, the independent variable was once again the participant's status (i.e., unaffected and pwMS with low-mild or moderate-severe disability), while the dependent variables were the 3 previously listed symmetry indexes calculated at hip, knee and ankle joints. Two additional analyses were carried out by pooling all the pwMS in a single group.The level of significance was set at $p = 0.05$, and the effect sizes were assessed using the eta-squared (η^2) coefficient.

Univariate ANOVA was carried out as a post hoc test by reducing the level of significance to $p = 0.007$ (0.05/7) for spatio-temporal parameters and $p = 0.017$ (0.05/3) for the symmetry indexes after a Bonferroni correction for multiple comparisons. When necessary, a post hoc Holm-Sidak test for pairwise comparison was carried out to assess intra- and inter-group differences. Data were checked for normality (using the Shapiro–Wilk test) and homogeneity of variances (by means of Levene's test) before any ANOVA.

Moreover, for the group of pwMS only, we also explored the existence of a relationship between gait symmetry parameters, spatio-temporal parameters of gait and disability level using Spearman's rank correlation coefficient rho by setting the level of significance at $p = 0.05$. Rho values of 0.1, 0.3 and 0.5 were assumed to be representative of small, moderate, and large correlations, respectively, according to Cohen's guidelines [45]. All analyses were performed using the IBM SPSS Statistics v.20 software (IBM, Armonk, NY, USA).

3. Results

The results of the comparison between pwMS and unaffected individuals as regards spatio-temporal parameters of gait and symmetry indexes are summarized in Tables 2 and 3.

Table 2. Comparison between spatio-temporal parameters of gait of people with MS and unaffected individuals. Stance, swing and double support phases duration are expressed as percentage of the gait cycle. Values are expressed as mean (SD).

	Healthy Controls	All MS	MS Low-Mild Disability (EDSS $\leq$3.5)	MS Moderate-Severe Disability (EDSS >3.5)
Gait Speed (m/s)	1.23 (0.19)	0.85 (0.34) [a]	1.00 (0.31) [a]	0.65 (0.27) [a,b]
Stride Length (m)	1.29 (0.13)	1.02 (0.25) [a]	1.09 (0.22) [a]	0.92 (0.24) [a,b]
Cadence (steps/min)	113.07 (10.34)	96.49 (20.26) [a]	104.48 (17.06) [a]	85.26 (19.17) [a,b]
Step Width (m)	0.20 (0.03)	0.22 (0.04) [a]	0.21 (0.03)	0.23 (0.04) [a]
Stance Phase	59.09 (2.80)	63.63 (4.82) [a]	62.51 (4.03) [a]	65.22 (5.41) [a,b]
Swing Phase	40.45 (1.76)	35.78 (4.78) [a]	37.22 (4.01) [a]	33.75 (5.08) [a,b]
Double Support	19.86 (3.60)	29.38 (10.72) [a]	25.58 (8.24) [a]	34.7 (11.60) [a,b]

The symbol [a] indicates significant difference vs. Healthy Controls after Bonferroni correction. The symbol [b] indicates significant difference vs. people with MS with low-mild disability after Bonferroni correction.

Table 3. Comparison between symmetry indexes of people with MS and unaffected individuals. Values are expressed as mean (SD).

Cyclogram Parameter		Healthy Controls	All MS	MS Low-Mild Disability (EDSS ≤ 3.5)	MS Moderate-Severe Disability (EDSS >3.5)
Area	**Hip**	108.17 (98.54)	195.52 (190.40) [a]	144.16 (163.77)	267.68 (203.34) [a,b]
Orientation ϕ		1.58 (1.34)	5.09 (7.06) [a]	2.28 (2.64)	9.05 (9.18) [a,b]
Trend Symmetry		0.26 (0.43)	1.74 (2.97) a	0.66 (0.98)	3.27 (4.02) [a,b]
Area	**Knee**	270.60 (192.50)	311.18 (269.64)	262.71 (259.20)	379.28 (272.33)
Orientation ϕ		1.51 (1.57)	6.99 (9.60) [a]	2.22 (2.46)	13.71 (11.71) [a,b]
Trend Symmetry		0.48 (0.41)	4.19 (6.89) [a]	1.26 (2.01)	8.29 (8.98) [a,b]
Area	**Ankle**	76.45 (62.25)	91.07 (82.40)	74.50 (68.90)	114.33 (94.33)
Orientation ϕ		3.05 (2.80)	6.45 (6.48) [a]	4.88 (5.38) [a]	8.65 (7.28) [a,b]
Trend Symmetry		1.51 (1.58)	8.46 (10.00) [a]	5.40 (9.70) [a]	12.77 (8.84) [a,b]

The symbol [a] indicates significant difference vs. Healthy Controls after Bonferroni correction. The symbol [b] indicates significant difference vs. people with MS with low-mild disability after Bonferroni correction.

3.1. *Spatio-Temporal Parameters of Gait*

Parameters that were separately calculated for right and left limb (i.e., stride length and duration of stance, swing and double support phases) were preliminarily screened using an independent sample t-test to verify the existence of significant differences between the limbs. Since this was not the case, their average value was calculated and considered representative of a certain participant.

The statistical analysis revealed a significant effect of the individual's status (F (14,346) = 10.70, $p < 0.001$, Wilks $\lambda = 0.49$, $\eta^2 = 0.30$) on spatio-temporal parameters of gait. In particular, the follow-up analysis detected the existence of significant differences between the three groups in all the parameters investigated except for step width. In this case, no significant differences were found between unaffected individuals and pwMS with low–mild disability, while those with moderate–severe disability exhibited a step width significantly higher with respect to healthy controls (0.23 m vs. 0.20 m, $p = 0.007$).

3.2. *Gait Symmetry Indexes*

MANOVA detected a significant effect of the individual's status on symmetry indexes in all three joints investigated. In particular, for hip [F (6354) = 13.48, $p < 0.001$, Wilks $\lambda = 0.66$, $\eta^2 = 0.19$], for knee (F (6354) = 21.35, $p < 0.001$, Wilks $\lambda = 0.54$, $\eta^2 = 0.27$) and for ankle (F (6354) = 12.28, $p < 0.001$, Wilks $\lambda = 0.68$, $\eta^2 = 0.17$). From the post hoc analysis, it was observed that in the case of cyclogram area no significant differences were observed between the groups as regards knee and ankle joints, while in the case of the hip joint, pwMS with moderate–severe disability exhibited significantly larger areas in comparison with both unaffected individuals and pwMS with low–mild disability. The orientation and Trend Symmetry indexes were found significantly different in the three groups at the ankle joint. In the case of hip and knee, significant differences were observed between the moderate–severe disability group with both low–mild disability and unaffected individual groups. Figure 2 shows an example of the different shapes and orientations of the cyclograms for pwMS of different disability levels and unaffected individuals.

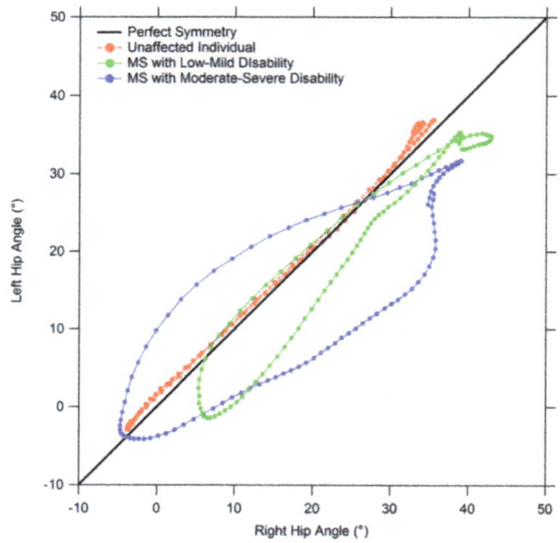

Figure 2. Example of comparison between cyclograms of unaffected individuals and people with MS of different disability levels. The diagram refers to the hip joint.

3.3. *Relationship between Symmetry Indexes and Spatio-Temporal Parameters of Gait*

Table 4 summarizes the results of the correlation analysis between disability level, spatio-temporal parameters of gait and symmetry indexes for pwMS. Significant correlations were found between all the variables investigated, with a few exceptions, which involved the cyclograms' area. For this parameter, we generally observed the weakest associations with spatio-temporal parameters of gait or (as in the case of the ankle joint) no correlations at all, except for a low one with EDSS score. Instead, Trend Symmetry was the index that exhibited the largest coefficient of correlation with EDSS score (rho ranged from 0.62 to 0.69 depending on the joint: $p < 0.001$), gait speed (−0.58 to −0.63, $p < 0.001$), stride length (−0.52 to −0.55, $p < 0.001$) and double support phase duration (0.50 to 0.57, $p < 0.001$) in all three joints.

Table 4. Spearman's coefficients for the correlations between spatio-temporal parameters of gait, symmetry indexes and disability level in people with MS.

		EDSS Score	Speed	Stride Length	Cadence	Step width	Double Support
Area	**Hip**	0.433 **	−0.268 **	−0.245 *	−0.240 *	0.147	0.321 **
Orientation ϕ	**Hip**	0.509 **	−0.511 **	−0.475 **	−0.403 **	0.262 **	0.426 **
Trend Symmetry	**Hip**	0.619 **	−0.581 **	−0.519 **	−0.493 **	0.349 **	0.568 **
Area	**Knee**	0.314 **	−0.225 *	−0.322 **	−0.093	0.318 **	0.228 *
Orientation ϕ	**Knee**	0.644 **	−0.590 **	−0.486 **	−0.517 **	0.473 **	0.524 **
Trend Symmetry	**Knee**	0.687 **	−0.634 **	−0.546 **	−0.547 **	0.419 **	0.532 **
Area	**Ankle**	0.223 *	−0.136	−0.046	−0.114	0.124	0.100
Orientation ϕ	**Ankle**	0.391 **	−0.439 **	−0.354 **	−0.464 **	0.281 **	0.376 **
Trend Symmetry	**Ankle**	0.636 **	−0.627 **	−0.512 **	−0.573 **	0.465 **	0.509 **

* $p < 0.05$; ** $p < 0.01$; EDSS: Expanded Disability Status Scale.

4. Discussion

4.1. *General Considerations*

The general aim of this study was to assess the magnitude of interlimb asymmetry during gait in pwMS in terms of joint kinematics and compare it with those of unaffected individuals using waveform-based methods. Such information is of great importance in the analysis of motor dysfunctions associated with MS because symmetry has been recognized as one of the domains that significantly influences gait quality and efficiency [46], together with pace, rhythm, variability and complexity. Although the analysis based on cyclograms is somewhat complex and requires full kinematic data (which can be typically extracted only from laboratory tests) it may provide a better insight into the mechanisms that lead to altered gait in pwMS. Moreover, it represents an effective way to quantify the deviation from a "normal" gait through parameters easy to interpret and thus may be useful, for instance, to quickly verify the effects of rehabilitation, or training exercise, on a joint-by-joint basis.

Our results show that pwMS exhibit a significantly larger asymmetry with respect to unaffected participants for hip, knee and ankle joints when considering cyclogram orientation and the Trend Symmetry parameter. In contrast, a more conventional parameter such as the cyclogram area was able to discriminate pwMS from controls only as regards the hip joint. The approach employed also appears capable of detecting asymmetry differences associated with the disability level of pwMS since all investigated parameters (with the same exceptions involving the cyclogram areas) were found significantly higher in pwMS with moderate–severe disability with respect to those with low–mild disability. In contrast, the analysis generally failed in discriminating pwMS with EDSS $\leq$3.5 from unaffected individuals, even though some significant differences were observed at the ankle joint as regards the cyclogram orientation and Trend Symmetry. Such phenomenon suggests that the ankle joint might play a specific role in gait alterations. This seems also confirmed by recent studies [47], which reported greater ankle muscle coactivation (with respect to unaffected individuals) and alterations in ankle joint kinematics during gait occurring especially at early stages of the disease in pwMS and that might serve as biomarker of neurodegeneration. It is also possible that the absence of significant differences between the two groups of pwMS depends on the specific EDSS score cut-off selected to stratify the participants. Further studies are thus necessary to clarify such aspects.

On one hand, the findings of the present study confirm those reported by Crenshaw et al. [30] for a small cohort of pwMS, but further extend them, as they indicate that asymmetry tends to be more marked as the disability level worsens. Generally speaking, several previous studies on gait of pwMS included some form of asymmetry analysis, but this is often restricted to few spatio-temporal parameters. In this regard, there is strong evidence that pwMS exhibit clinically relevant asymmetries in terms of gait cycle duration, stride/step length and time and stance/swing phase duration [21,48–50]. To the authors' knowledge, only two studies [17,51] specifically investigated asymmetry for lower limb joint kinematics during gait, even when using discrete values of ROM (typically the maximum value observed within the entire gait cycle). Consistent with our results, they both reported higher asymmetries in pwMS with respect to unaffected individuals at hip, knee and ankle joints.

The existence of interlimb asymmetry in terms of joint kinematics can be attributed to several factors. Firstly, the differences in muscular function, due to corticospinal tracts involvement, between more affected and less affected limb (which has been repeatedly observed in pwMS in terms of strength, torque and metabolism [9]) may introduce some kind of unbalance even on joint movement control. This can be further exacerbated by the presence of compensatory mechanisms unconsciously adopted to overcome the uneven supporting and propulsive action of the two limbs. Secondly, the reduced capability to optimally coordinate left and right limbs during gait might be due to reduced efficiency in the neural communication pathways between the two cerebral hemispheres, particularly as regards the fiber bundle connecting the primary motor cortices [52]. Moreover, imaging studies have highlighted the existence of a significant correlation between asymmetries in electrophysiological deficits for both arms and legs and asymmetric anatomic changes in the spinal cord's normal-appearing white matter, thus suggesting that the functional asymmetries are associated with microstructural damage of the spinal cord [53]. Finally, Filli et al. [17] hypothesized that the loss of inter- and intralimb coordination, particularly at the distal level, might be due to the altered integrity of the long ascending and descending myelinated fiber tracts of cortical, cerebellar and brainstem systems.

4.2. Relationship between Interlimb Asymmetry, Spatio-Temporal Parameters of Gait and Disability

As previously mentioned, one of the most debated issues related to lower limb asymmetry in MS involves assessment of its actual impact on gait performance. In this regard, the findings of the present study suggest the existence of a close relationship between gait efficiency and interlimb asymmetry of joint kinematics, especially when the latter is expressed in terms of cyclogram orientation and Trend Symmetry. This link appears similar in strength regardless of the joint considered for gait speed and stride length (and consequently for cadence). However, we also detected moderate to large correlations between asymmetry parameters and other aspects of gait more specifically associated with dynamic balance, such as step width and double support phase duration.

The recent reviews by Rudroff and Proessl [9] and Ramari et al. [54], which analyzed the effect of asymmetries in muscular strength and limb loading on walking capabilities of pwMS (in particular gait speed and performance on timed tests) raised strong doubts about the possibility of defining a clear relationship between them. However, recent studies that investigated asymmetry through calculation of the phase relationship between the step timing of the left and right legs (the so-called Phase Coordination Index, PCI [55]) reported that bilateral coordination of gait was negatively correlated with gait speed and performance in 6 m and Timed 25-foot walking tests [52,56]. Even from a quantitative point of view, such results are fully consistent with those of the present study, thus suggesting that even when assessed with completely independent methods, bilateral coordination negatively affects gait speed and stride length [49]. Interestingly, we also observed significant positive correlations between symmetry parameters and step width and double support, the latter being stronger. Although there are no data available for comparison, it has been suggested that in pwMS, asymmetries in muscle strength may result in a wider base of support and prolonged double support phase duration during gait [15]. Although in this study we did not investigate muscular strength, it appears reasonable to hypothesize that even the asymmetry in kinematics of lower limb joints (through a combined or superposed effect with those of muscle function) plays a crucial role in the establishment of adaptative strategies that pwMS are forced to employ to counteract the negative effects associated with uneven motor functions of the two limbs.

Finally, it is to be mentioned that all asymmetry parameters were found positively correlated with the EDSS score, thus indicating the strict relationship existing between bilateral coordination and disease progression, whose nature deserves further in-depth investigations. This result was not completely surprising, since gait deterioration represents one of the distinctive hallmarks of the disease, but it is noteworthy that similar findings were also found by Plotnik et al. [56], who, as previously mentioned, calculated a different index of asymmetry (i.e., the previously mentioned PCI).

Some limitations of the study are to be acknowledged. Firstly, in our research, the waveform-based method was employed only to explore interlimb symmetry, but the same approach might be advantageously exploited to investigate intralimb coordination considering the different combination of joints (i.e., hip vs. knee, knee vs. ankle, etc.). This would provide further important data regarding the possible impact of the degree of coordination (or incoordination) between the two limbs on the quality of coordination between the joints and vice versa, thus allowing assessment of the existence and type of compensatory mechanisms. Furthermore, in the present study, men and women were pooled in a single group, even though recent studies point out that several sex-related differences exist in lower limb kinematics during gait for pwMS [57]. At last, it should be considered that all the walking tests performed for the present study refer to a relatively short distance (i.e., 10 m), but the literature reports that, in pwMS, asymmetry of gait (calculated in terms of spatio-temporal parameters of gait) tends to worsen in case of longer distance due to fatigue effects [48–50]. It would be, thus, interesting to verify if a similar phenomenon would be present also as regards the joint kinematics symmetry.

5. Conclusions

The analysis of asymmetry of lower limb joint kinematics during gait of pwMS shows that bilateral coordination is impaired in those with moderate–severe disability at hip, knee and ankle levels, while individuals characterized by low–mild disability exhibit anomalous values of asymmetry at ankle level only. Moreover, the existence of moderate-to-large correlations between symmetry and gait parameters suggest that the former (which increases as the disease progresses) has a direct influence on gait quality and efficiency since pwMS with the poorest symmetry indexes are characterized by reduced gait speed and stride length and increased step width and double support phase duration. While confirming that MS differentially alters most aspects of lower limb motor functionality, the findings of the present study also suggest that the asymmetries of spatio-temporal parameters reported by many studies on gait of pwMS are likely to reflect the combined effect of muscular and joint kinematics asymmetries. In such a context, the use of waveform-based methods to assess interlimb (and possibly interlimb) symmetry may provide useful insights not only to better understand the impairments in motor control associated with the presence of MS, but also to accurately assess the effect of physical therapy and exercise training programs, which have been shown to have a positive effect on gait and balance asymmetries of individuals with MS as well as other chronic neurologic conditions [58]. However, future studies (possible longitudinal) are necessary to clarify the evolution of asymmetry during the disease progression, to identify specific peculiarities associated with MS type and with the sex of the affected individual and to assess the effects of fatigue.

Author Contributions: Conceptualization, M.P. (Massimiliano Pau); methodology, M.P. (Massimiliano Pau), B.L.; software, B.L.; validation, B.L., M.P. (Massimiliano Pau); formal analysis, M.D., F.P.; investigation, E.C., G.C., M.P. (Micaela Porta); resources, E.C.; data curation, M.D., M.P. (Massimiliano Pau); writing—original draft preparation, M.P. (Massimiliano Pau); writing—review and editing, E.C.; supervision, E.C.; project administration, E.C.; funding acquisition, E.C. All authors have read and agreed to the published version of the manuscript.

Funding: This research was funded by the Autonomous Region of Sardinia (L.R. 7/2007, grant number RASSR42584).

Institutional Review Board Statement: Data used for this retrospective study were collected during previous studies conducted according to the guidelines of the Declaration of Helsinki, and approved by the Local Ethics Committee (authorization numbers 180/2014, 102/2018 and 198/2019).

Informed Consent Statement: Informed consent was obtained from all subjects involved in the study.

Data Availability Statement: The data that support the findings of this study are available from the corresponding author M.P. upon reasonable request.

Acknowledgments: The authors are grateful to Federico Arippa, Federica Corona and Giuseppina Pilloni for their valuable support during the data acquisition process.

Conflicts of Interest: The authors declare no conflict of interest.

Appendix A

Matlab pseudocode for the calculation of symmetry parameters (see ref. [44] for details)

```
function [TS, CO, m_T] = symm_pars(X, Y)
% symm_pars calculates inter-limb joint cyclogram symmetry parameters.
% input:
% X = column array of Left joint data.
% Y = column array of Right joint data.
% Output:
% 1 - TS = Trend Symmetry, as defined by Crenshaw et al.(2006) [44]
% 2 - CO = Cyclogram Orientation (degrees).
```

```
    % 3 - m_T = Angular coefficient of the trend line.
    XT = X - mean(X);
    YT = Y - mean(Y);
    M = [XT YT];
    S = (M')*M;
    [V, D, W] = eig(S); % eig function returns full matrix W whose columns
are the corresponding left eigenvectors, so that W'*A = D*W'.
    eigVals = sum(D); % array containing the eigenvalues or Inertia matrix
    [emax, pos_emax] = max(eigVals); % emax = maximm eigenvalue (i.e.
maximum variability; pos_emax = position of emax in array eigVals;
    [emin, pos_emin] = min(eigVals); % emin = minimum eigenvalue (i.e.
minimum variability; pos_emax = position of emax in array eigVals;
    e1 = V(:,pos_emax); % eigenvector parallel to the direction maximizing
the variability (along which the variability is maximum)
    % NOTE:
    % from the mathematical point of view, the eigenvalues and the
eigenvectors
    % of matrix M represent, respectively, the principal inertia moments
and
    % the direction of principal axis of inertia of the cyclograms point
distribution
    TS = (emin/emax)*100; % Trend Symmetry:  the ratio of the minimum to
the maximum variability expressed as percentage;
    % in condition of perfect symmetry, the direction of e2 is 45° with
respect
    % to the reference axis.  "delta_THETA_I", i.e.  the difference between
the orientation of e1 and
    % 45°, is a measurement of the asymmetry of the cyclogram points
    CO = 45 - (180/pi)*atan(e1(2)/e1(1)); % angle between the eigenvector e2
and 45 degrees
    m_T = (e1(2)/e1(1)); % Angular coefficient of the trend line
    Fig1 = figure;
    p1 = plot(XT, YT, 'or'); %Cyclogram
    axis equal
    grid on
    hold on
    p2 = plot(XT, XT, '-k'); % 45° line
    p3 = plot(XT, m_T.*XT, '-r'); % principal axis
    legend('cyclogram', '45° line', 'linear regression', 'principal axis');
    xlabel('left joint (deg)', 'fontsize', 5);
    ylabel('right joint (deg)', 'fontsize', 5);
    title('syncronized cyclogram');
    end
```

References

1. Compston, A.; Coles, A. Multiple sclerosis. *Lancet* **2008**, *372*, 1502–1517. [CrossRef]
2. Koch-Henriksen, N.; Sørensen, P.S. The changing demographic pattern of multiple sclerosis epidemiology. *Lancet Neurol.* **2010**, *9*, 520–532. [CrossRef]
3. Thompson, A.J.; Baranzini, S.E.; Geurts, J.; Hemmer, B.; Ciccarelli, O. Multiple sclerosis. *Lancet* **2018**, *91*, 1622–1636. [CrossRef]
4. Cameron, M.H.; Lord, S. Postural control in multiple sclerosis: Implications for fall prevention. *Curr. Neurol. Neurosci. Rep.* **2010**, *10*, 407–412. [CrossRef] [PubMed]
5. Comber, L.; Galvin, R.; Coote, S. Gait deficits in people with multiple sclerosis: A systematic review and meta-analysis. *Gait Posture* **2017**, *51*, 25–35. [CrossRef]
6. Lamers, I.; Feys, P. Assessing upper limb function in multiple sclerosis. *Mult. Scler.* **2014**, *20*, 775–784. [CrossRef]

7. Buzaid, A.; Dodge, M.P.; Handmacher, L.; Kiltz, P.J. Activities of daily living: Evaluation and treatment in persons with multiple sclerosis. *Phys. Med. Rehabil Clin. N. Am.* **2013**, *24*, 629–638. [CrossRef]
8. Institute of Medicine. *Multiple Sclerosis: Current Status and Strategies for the Future*; The National Academies Press: Washington, DC, USA, 2001.
9. Rudroff, T.; Proessl, F. Effects of Muscle Function and Limb Loading Asymmetries on Gait and Balance in People with Multiple Sclerosis. *Front. Physiol.* **2018**, *9*, 531. [CrossRef]
10. Flachenecker, P.; Henze, T.; Zettl, U.K. Spasticity in patients with multiple sclerosis: Clinical characteristics, treatment and quality of life. *Acta Neurol. Scand.* **2014**, *129*, 154–162. [CrossRef] [PubMed]
11. Chung, L.H.; Remelius, J.G.; van Emmerik, R.E.; Kent-Braun, J.A. Leg power asymmetry and postural control in women with multiple sclerosis. *Med. Sci. Sports Exerc.* **2008**, *40*, 1717–1724. [CrossRef]
12. Larson, R.D.; McCully, K.K.; Larson, D.J.; Pryor, W.M.; White, L.J. Bilateral differences in lower-limb performance in individuals with multiple sclerosis. *J. Rehabil. Res. Dev.* **2013**, *50*, 215–222. [CrossRef]
13. Proessl, F.; Ketelhut, N.B.; Rudroff, T. No association of leg strength asymmetry with walking ability, fatigability, and fatigue in multiple sclerosis. *Int. J. Rehabil. Res.* **2018**, *41*, 267–269. [CrossRef] [PubMed]
14. Farrell III, J.W.; Motl, R.W.; Learmonth, Y.C.; Pilutti, L.A. Persons with Multiple Sclerosis Exhibit Strength Asymmetries in both Upper and Lower Extremities. *Physiotherapy* **2020**, *25*, S0031-9406(20)30392-8. [CrossRef] [PubMed]
15. Workman, C.D.; Fietsam, A.C.; Rudroff, T. Associations of lower limb joint asymmetry with fatigue and disability in people with multiple sclerosis. *Clin. Biomech.* **2020**, *75*, 104989. [CrossRef] [PubMed]
16. Daunoraviciene, K.; Ziziene, J.; Ovcinikova, A.; Kizlaitiene, R.; Griskevicius, J. Quantitative body symmetry assessment during neurological examination. *Technol. Health Care* **2020**, *28*, 573–584. [CrossRef]
17. Filli, L.; Sutter, T.; Easthope, C.S.; Killeen, T.; Meyer, C.; Reuter, K.; Lörincz, L.; Bolliger, M.; Weller, M.; Curt, A.; et al. Profiling walking dysfunction in multiple sclerosis: Characterisation, classification and progression over time. *Sci. Rep.* **2018**, *8*, 4984. [CrossRef] [PubMed]
18. Cofré-Lizama, L.E.; Khan, F.; Lee, P.V. The use of laboratory gait analysis for understanding gait deterioration in people with multiple sclerosis. *Mult. Scler.* **2016**, *22*, 1768–1776. [CrossRef]
19. Robinson, R.O.; Herzog, W.; Nigg, B.M. Use of force platform variables to quantify the effects of chiropractic manipulation on gait symmetry. *J. Manip. Physiol. Ther.* **1987**, *10*, 172–176.
20. Viteckova, S.; Kutilek, P.; Svoboda, Z.; Krupicka, R.; Kauler, J.; Szabo, Z. Gait symmetry measures: A review of current and prospective methods. *Biomed. Signal Proces.* **2018**, *42*, 89–100. [CrossRef]
21. Dujmovic, I.; Radovanovic, S.; Martinovic, V.; Dackovic, J.; Maric, G.; Mesaros, S.; Pekmezovic, T.; Kostic, V.; Drulovic, J. Gait pattern in patients with different multiple sclerosis phenotypes. *Mult. Scler. Relat. Disord.* **2017**, *13*, 13–20. [CrossRef]
22. Moraes, A.G.; Neri, S.G.R.; Motl, R.W.; Tauil, C.B.; Glehn, F.V.; Corrêa, É.C.; de David, A.C. Effect of hippotherapy on walking performance and gait parameters in people with multiple sclerosis. *Mult. Scler. Relat. Disord.* **2020**, *43*, 102203. [CrossRef]
23. Liao, F.; Wang, J.; He, P. Multi-resolution entropy analysis of gait symmetry in neurological degenerative diseases and amyotrophic lateral sclerosis. *Med. Eng. Phys.* **2008**, *30*, 299–310. [CrossRef] [PubMed]
24. Xie, H.-B.; Zheng, Y.-P.; Guo, J.-Y.; Chen, X. Cross-fuzzy entropy: A new method to test pattern synchrony of bivariate time series. *Inf. Sci.* **2010**, *180*, 1715–1724. [CrossRef]
25. Kutilek, P.; Viteckova, S.; Svoboda, Z.; Socha, V.; Smrcka, P. Kinematic quantification of gait asymmetry based on characteristics of angle-angle diagrams. *Acta Polytechnica Hungarica* **2014**, *11*, 25–38.
26. Viteckova, S.; Kutilek, P.; Krupicka, R.; Adamova, B.; Szabo, Z.; de Brito, A.C.D.M.; Kopecka, J. Evaluation of movement of patients with Parkinson's disease using wearable MoCap system and bilateral cyclograms. In *Proceeding of the International Conference on Applied Electronics, Plisen, Chech Republic, 5–6 September 2017*; University of West Bohami: Plisen, Chech Republic, 2017; pp. 269–272. [CrossRef]
27. Pilkar, R.; Ramanujam, A.; Chervin, K.; Forrest, G.; Nolan, K.J. Cyclogram based Joint Symmetry Assessment after Utilization of a Foot Drop Stimulator during Post Stroke Hemiplegic Gait. *J. Biomech. Eng.* **2018**. [CrossRef] [PubMed]
28. Sung, P.S.; Danial, P. A Kinematic Symmetry Index of Gait Patterns Between Older Adults with and Without Low Back Pain. *Spine* **2017**, *42*, E1350–E1356. [CrossRef]
29. Farkas, G.J.; Schlink, B.R.; Fogg, L.F.; Foucher, K.C.; Wimmer, M.A.; Shakoor, N. Gait asymmetries in unilateral symptomatic hip osteoarthritis and their association with radiographic severity and pain. *Hip. Int.* **2019**, *29*, 209–214. [CrossRef]
30. Bai, X.; Ewins, D.; Crocombe, A.D.; Xu, W. Kinematic and biomimetic assessment of a hydraulic ankle/foot in level ground and camber walking. *PLoS ONE* **2017**, *12*, e0180836. [CrossRef] [PubMed]
31. Grieve, D.W. Gait patterns and the speed of walking. *Biomed. Eng.* **1968**, *3*, 119–122.
32. Crenshaw, S.J.; Richards, J.G.; Miller, C.M. Gait Symmetry in Subjects with Multiple Sclerosis. *Med. Sci. Sports Exerc.* **2006**, *38*, S1. [CrossRef]
33. Polman, C.H.; Reingold, S.C.; Edan, G.; Filippi, M.; Hartung, H.P.; Kappos, L.; Lublin, F.D.; Metz, L.M.; McFarland, H.F. Diagnostic criteria for multiple sclerosis: 2005 revisions to the McDonald criteria. *Ann. Neurol.* **2005**, *58*, 840–846. [CrossRef] [PubMed]
34. Polman, C.H.; Reingold, S.C.; Banwell, B.; Clanet, M.; Cohen, J.A.; Filippi, M.; Fujihara, K.; Havrdova, E.; Hutchinson, M.; Kappos, L.; et al. Diagnostic criteria for multiple sclerosis: 2010 revisions to the McDonald criteria. *Ann. Neurol.* **2011**, *69*, 292–302. [CrossRef]

35. Pau, M.; Coghe, G.; Atzeni, C.; Corona, F.; Pilloni, G.; Marrosu, M.G.; Cocco, E.; Galli, M. Novel characterization of gait impairments in people with multiple sclerosis by means of the gait profile score. *J. Neurol. Sci.* **2014**, *345*, 159–163. [CrossRef] [PubMed]
36. Pau, M.; Coghe, G.; Corona, F.; Marrosu, M.G.; Cocco, E. Effect of spasticity on kinematics of gait and muscular activation in people with Multiple Sclerosis. *J. Neurol. Sci.* **2015**, *358*, 339–344. [CrossRef] [PubMed]
37. Pau, M.; Corona, F.; Coghe, G.; Marrosu, M.G.; Cocco, E. Quantitative assessment of the effects of 6 months of adapted physical activity on gait in people with multiple sclerosis: A randomized controlled trial. *Disabil. Rehabil.* **2017**, *13*, 1–11. [CrossRef] [PubMed]
38. Wells, R.P. The kinematics and energy variations of swing-through crutch gait. *J. Biomech.* **1979**, *12*, 579–585. [CrossRef]
39. Crosbie, J. Kinematics of walking frame ambulation. *Clin. Biomech.* **1993**, *8*, 31–36. [CrossRef]
40. Davis, R.B.; Õunpuu, S.; Tyburski, D. A gait analysis data collection and reduction technique. *Hum. Mov. Sci.* **1991**, *10*, 575–587. [CrossRef]
41. Goswami, A. Kinematic quantification of gait symmetry based on bilateral cyclograms. In Proceedings of the International Society of Biomechanics XIXth Congress, Dunedin, New Zealand, 6–11 July 2003.
42. Hershler, C.; Milner, M. Angle-angle diagrams in the assessment of locomotion. *Am. J. Phys. Med.* **1980**, *59*, 109–125.
43. Goswami, A. A new gait parameterization technique by means of cyclogram moments: Application to human slope walking. *Gait Posture* **1998**, *8*, 15–36. [CrossRef]
44. Crenshaw, S.J.; Richards, J.G. A method for analyzing joint symmetry and normalcy, with an application to analyzing gait. *Gait Posture* **2006**, *24*, 515–521. [CrossRef]
45. Cohen, J. *Statistical Power Analysis for the Behavioral Sciences*, 2nd ed.; Lawrence Erlbaum Associates: Hillsdale, MI, USA, 1998.
46. Lord, S.; Galna, B.; Rochester, L. Moving forward on gait measurement: Toward a more refined approach. *Mov. Disord.* **2013**, *28*, 1534–1543. [CrossRef] [PubMed]
47. Cofré Lizama, L.E.; Bastani, A.; van der Walt, A.; Kilpatrick, T.; Khan, F.; Galea, M.P. Increased ankle muscle coactivation in the early stages of multiple sclerosis. *Mult. Scler. J. Exp. Transl. Clin.* **2020**, *11*, 6. [CrossRef] [PubMed]
48. Kalron, A. Association between perceived fatigue and gait parameters measured by an instrumented treadmill in people with multiple sclerosis: A cross-sectional study. *J. Neuroeng. Rehabil.* **2015**, *12*, 34. [CrossRef]
49. Escudero-Uribe, S.; Hochsprung, A.; Izquierdo-Ayuso, G. Gait pattern changes after six-minute walk test in persons with multiple sclerosis. *Physiother. Res. Int.* **2019**, *24*, e1741. [CrossRef] [PubMed]
50. Shema-Shiratzky, S.; Gazit, E.; Sun, R.; Regev, K.; Karni, A.; Sosnoff, J.J.; Herman, T.; Mirelman, A.; Hausdorff, J.M. Deterioration of specific aspects of gait during the instrumented 6-min walk test among people with multiple sclerosis. *J. Neurol.* **2019**, *266*, 3022–3030. [CrossRef]
51. Taborri, J.; Studer, V.; Grossi, P.; Brambilla, L.; Patanè, F.; Ferrò, M.T.; Mantegazza, R.; Rossi, S. Reliability and Repeatability Analysis of Indices to Measure Gait Deterioration in MS Patients during Prolonged Walking. *Sensors* **2020**, *20*, 5063. [CrossRef]
52. Richmond, S.B.; Swanson, C.W.; Peterson, D.S.; Fling, B.W. A temporal analysis of bilateral gait coordination in people with multiple sclerosis. *Mult. Scler. Relat. Disord.* **2020**, *45*, 102445. [CrossRef]
53. Von Meyenburg, J.; Wilm, B.J.; Weck, A.; Petersen, J.; Gallus, E.; Mathys, J.; Schaetzle, E.; Schubert, M.; Boesiger, P.; von Meyenburg, K.; et al. Spinal cord diffusion-tensor imaging and motor-evoked potentials in multiple sclerosis patients: Microstructural and functional asymmetry. *Radiology* **2013**, *267*, 869–879. [CrossRef]
54. Ramari, C.; Hvid, L.G.; David, A.C.; Dalgas, U. The importance of lower-extremity muscle strength for lower-limb functional capacity in multiple sclerosis: Systematic review. *Ann. Phys. Rehabil. Med.* **2020**, *63*, 123–137. [CrossRef]
55. Plotnik, M.; Giladi, N.; Hausdorff, J.M. A new measure for quantifying the bilateral coordination of human gait: Effects of aging and Parkinson's disease. *Exp. Brain Res.* **2007**, *181*, 561–570. [CrossRef]
56. Plotnik, M.; Wagner, J.M.; Adusumilli, G.; Gottlieb, A.; Naismith, R.T. Gait asymmetry, and bilateral coordination of gait during a six-minute walk test in persons with multiple sclerosis. *Sci. Rep.* **2020**, *10*, 12382. [CrossRef] [PubMed]
57. Pau, M.; Corona, F.; Pilloni, G.; Porta, M.; Coghe, G.; Cocco, E. Do gait patterns differ in men and women with multiple sclerosis? *Mult. Scler. Relat. Disord.* **2017**, *18*, 202–208. [CrossRef] [PubMed]
58. Farrell, J.W., III; Merkas, J.; Pilutti, L.A. The Effect of Exercise Training on Gait, Balance, and Physical Fitness Asymmetries in Persons with Chronic Neurological Conditions: A Systematic Review of Randomized Controlled Trials. *Front. Physiol.* **2020**, *11*, 585765. [CrossRef] [PubMed]

symmetry

MDPI

Article

Effects of Limb Dominance on Postural Balance in Sportsmen Practicing Symmetric and Asymmetric Sports: A Pilot Study

Mohamed Abdelhafid Kadri [1,2], Frédéric Noé [1], Julien Maitre [1], Nicola Maffulli [3] and Thierry Paillard [1,*]

1. Laboratoire Mouvement, Equilibre, Performance et Santé, Université de Pau et des Pays de l'Adour, E2S UPPA, 11 rue Morane Saulnier, 65000 Tarbes, France; mohamed-abdelhafid.kadri1@uqac.ca (M.A.K.); frederic.noe@univ-pau.fr (F.N.); maitre.julien@univ-pau.fr (J.M.)
2. Laboratoire Etudes Sociales et Humaines et Analyse des Activités Physiques et Sportives, Département EPS, Université Badji Mokhtar Annaba, BP 12, Annaba 23000, Algeria
3. Centre for Sports and Exercise Medicine, Mile End Hospital, Barts and The London School of Medicine and Dentistry, London E1 4NS, UK; n.maffulli@qmul.ac.uk

* Correspondence: thierry.paillard@univ-pau.fr

Abstract: The current literature shows no consensus regarding the difference between the dominant leg (D-Leg) and the non-dominant leg (ND-Leg) in terms of postural control. This lack of consensus could stem from motor experience (i.e., symmetric or asymmetric motricity) and/or the physiological state induced by physical exercise. This study aimed to investigate the acute effects of fatiguing exercise on postural control when standing on the D-Leg and the ND-Leg, in athletes practicing symmetric (SYM) and asymmetric (ASYM) sports. Thirty healthy male participants were recruited and divided into two groups, (SYM n = 15) and (ASYM n = 15, on the basis of the motricity induced by the sport they practice. Monopedal postural control was assessed for the D-Leg and the ND-Leg before and after the fatigue period (which consisted of repeating squats until exhaustion). A force platform was used to calculate the spatio-temporal characteristics of the displacements of the center of foot pressure (COP). A significant fatigue effect was observed in both groups on the D-Leg and the ND-Leg for all the COP parameters. There was a tendency (p = 0.06) between the ASYM and SYM groups on the D-Leg, concerning the relative increase in the COP velocity in the frontal plane after the fatigue period. The fatigue condition disturbed postural control in both the SYM and ASYM groups on the D-Leg and ND-Leg. This disturbing effect related to fatigue tends to be more marked in athletes practicing asymmetric sports than in athletes practicing symmetric sports on the D-Leg.

Keywords: dominant leg; acute exercise; fatigue; sport practice

Citation: Kadri, M.A.; Noé, F.; Maitre, J.; Maffulli, N.; Paillard, T. Effects of Limb Dominance on Postural Balance in Sportsmen Practicing Symmetric and Asymmetric Sports: A Pilot Study. *Symmetry* **2021**, *13*, 2199. https://doi.org/10.3390/sym13112199

Academic Editor: John H. Graham

Received: 13 October 2021
Accepted: 14 November 2021
Published: 18 November 2021

1. Introduction

Leg dominance can be determined through the use of functional tests, such as ball kick, hop, or step up [1–3]. Most people use the dominant leg (D-Leg) to perform motor tasks, while using the non-dominant leg (ND-Leg) to support the body and stabilize posture [4]. Even though differences in terms of postural control have been reported between the D-Leg and the ND-Leg in athletes [5–8], other studies concluded that postural control was similar between the D-Leg and the ND-Leg among different athletes [9–13]. Paillard [14] hypothesized that this lack of consensus about the impact of limb dominance on monopedal postural control could stem from the nature of the sport practiced. With asymmetric activities that require frequent phases of monopedal posture on the ND-Leg to perform technical movements with the D-Leg (e.g., passing and kicking in soccer), the ND-Leg can display better postural control than the D-Leg [8,14,15]. In contrast, symmetric activities that use the two limbs similarly do not produce such an asymmetry of postural control [14]. Nevertheless, such a hypothesis still needs to be confirmed, since, to our knowledge, only two studies have been conducted to compare monopedal postural control in the D-Leg and the ND-Leg of expert athletes involved in asymmetric (ASYM) and symmetric (SYM) sports [6,15]. Moreover, the physiological state in which the subjects are evaluated can also act as a confounding factor by modulating the difference between the two legs [4,14]. Even though muscle fatigue negatively affects the perception of sensory information and control of the motor command of the postural system of both the D-Leg and ND-Leg [16,17], some studies performed with athletes showed that postural control was less affected by muscle fatigue on the ND-Leg than on the D-Leg [5,18], thus illustrating that the differences in postural control between the D-Leg and the ND-Leg could only be observed after the performance of a fatiguing exercise.

Hence, the present study was undertaken in order to accurately determine the effects of physiological states induced by fatiguing exercise on leg dominance in postural control, comparing athletes practicing SYM and ASYM sports. The latter were found to be more likely to be more sensitive to these acute effects [14]. It was hypothesized that, following the completion of fatiguing exercise, the dominant leg and the non-dominant leg could exhibit greater differences in postural control in athletes who participate in ASYM sports than those who participate in SYM sports.

2. Methods

Participants: Thirty healthy male athletes aged from 18 to 31 years old were recruited to participate in the experiment. They were divided into 2 groups, asymmetric (ASYM, n = 15) and symmetric (SYM, n = 15), based on the type of motricity in their sport practice, i.e., whether the movements carried out by the right and left side are symmetric or asymmetric. Age, morphological characteristics and details about sports participation for both ASYM and SYM groups are presented in Table 1. Volunteers likely to have musculoskeletal, vestibular, cardiovascular, ankle, knee or hip injuries in the last two years were excluded from the protocol. We asked the participants to avoid strenuous activity and not to eat or drink exciting substances 24 h prior to the data collection sessions. All participants gave their informed consent to participate in the experiment in accordance with the Declaration of Helsinki. The work has been approved by the Ethics Committee of the university authorities (14062014 Annaba).

Table 1. Participants' morphological characteristics expressed in median (IQR), sport practiced and expertise level of both ASYM and SYM groups.

	Groups	
Morphological Characteristics	**ASYM**	**SYM**
Age (years)	19 (5)	20 (7)
Height (cm)	175 (8)	174 (7)
Body weight (kg)	68.30 (10)	66.60 (10)
Body Mass Index (kg·m^{-2})	21.97 (2.28)	22.60 (2.57)
Foot size (cm)	28.38 (1.32)	28.38 (1.32)
Sport practiced	Soccer (n = 5) Handball (n = 2) Basketball (n = 1) Rugby (n = 1) Tennis (n = 3) Fencing (n = 2) Pelota (n = 1)	Track and field 800 m (n = 3) Track and field 1500 m (n = 1) Trail running (n = 2) Triathlon (n = 2) Biking (n = 4) Swimming (n = 3)
Sport competition level		
Local	n = 4	n = 3
Regional	n = 8	n = 9
National	n = 2	n = 2
International	n = 1	n = 1

Experimental Design: The experiment consisted of assessing postural control in a one-legged stance on the D-Leg and the ND-Leg (the dominant leg was defined as the leg used to kick a ball) for the ASYM and SYM groups in the following two conditions: (1) in an initial reference condition (REF condition), and (2) after a fatigue exercise (POST_FAT condition). In each condition and for each group, the D-Leg and ND-Leg were assessed in a counterbalanced order (Figure 1).

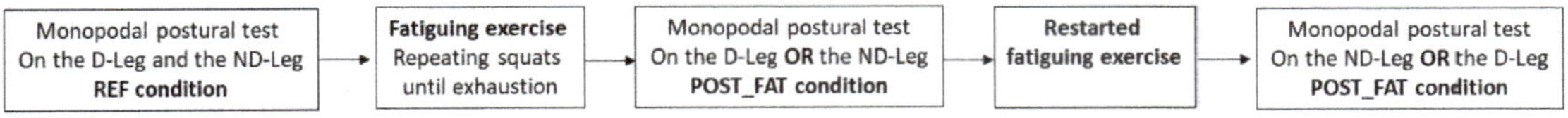

Figure 1. Chronological order of the protocol for both groups (ASYM and SYM).

Postural control assessment: Participants stood barefoot on a force platform (Stabilotest® Techno Concept, Mane, France; 40 Hz sampling frequency) that recorded the center of foot pressure (COP) displacements (spatio-temporal characteristics) with the PosturoWin v4 software. They were asked to sway and move as little as possible in a monopedal stance for 25 s with their arms alongside their body, while looking at a fixed target positioned 1 m in front of them at eye level. The unsupported leg (i.e., the free leg) was raised with a 90° joint flexion at the knee joint.

Reference condition: Participants performed 3 postural test trials on each leg with a 30 s rest between trials in order to achieve a stable postural score on monopedal stance and thus avoid learning effect between trials [19]. The third trial was recorded and corresponded to the REF condition.

Fatigue exercise: The fatigue exercise protocol consisted of repeating body-weight squats (i.e., without barbell) at a fixed and determined 0.5 Hz frequency (given by a metronome's sound beeps) until exhaustion, i.e., the inability to continue the squat exercise. Participants received verbal encouragements. The exercise had to be performed at 70° of knee flexion, which was determined with a goniometer (Comed®, Strasbourg, France). A rope was placed under the participants according to the 70° knee flexion angle and they were asked to touch it with their buttocks during each flexion in order to normalize the amplitude of the squat movements. A final postural control assessment was immediately performed at the end of the fatigue exercise on the D-Leg or the ND-Leg in POST_FAT condition in order to limit recovery, which can quickly impact postural control during the first 30 s following the exercise [17,20]. Since both legs could not be assessed consecutively during a 30 s period, the fatigue exercise was restarted before assessment of the following leg. The first and second durations of the fatigue exercise were recorded.

Data analysis: The COP surface area (90% confidence ellipse in mm^2) and mean COP velocity (sum of the cumulated COP displacement divided by the total time in $mm \cdot s^{-1}$) on the medio/lateral (frontal) and antero/posterior (sagittal) axes (COPx velocity and COPy velocity) were calculated as parameters that characterized postural control [21].

The relative increases between the REF and the POST_FAT conditions were calculated for all the parameters concerning the D-Leg and the ND-Leg as follows:

$$\text{POST\_FAT increase} = [(\text{POST\_FAT} - \text{REF}) \div \text{REF}] \times 100$$

Statistical analysis: Normality of the data was tested with the Shapiro–Wilk test. Non-parametric tests were used since the variables did not meet the assumption of normal distribution. Mann–Whitney tests for unpaired data were used to compare the morphological characteristics and the duration of the fatigue exercise protocol between the SYM and ASYM groups.

Paired samples Wilcoxon signed-rank tests were performed to compare COP parameters between the REF and POST_FAT conditions in order to determine a fatigue effect for the D-Leg and the ND-Leg in both ASYM and SYM groups. In order to test a potential effect of leg dominance, paired samples Wilcoxon signed-rank tests were performed within each group and in the two conditions to compare the values of COP parameters of the D-Leg and ND-Leg. Mann–Whitney tests for unpaired data were performed to compare the relative increases in all the COP parameters between the ASYM and SYM groups in order to determine a group effect. Results were considered significant at the level of 5%.

3. Results

Table 1 presents the morphological characteristics, sport specialty, and expertise level of the participants in the ASYM and SYM groups. Figure 2 presents the postural parameters of the D-Leg and the ND-Leg for the two groups in the two conditions. Figure 3 presents the relative increases in the postural parameters between the REF and POST_FAT conditions of the D-Leg and ND-Leg for the two groups.

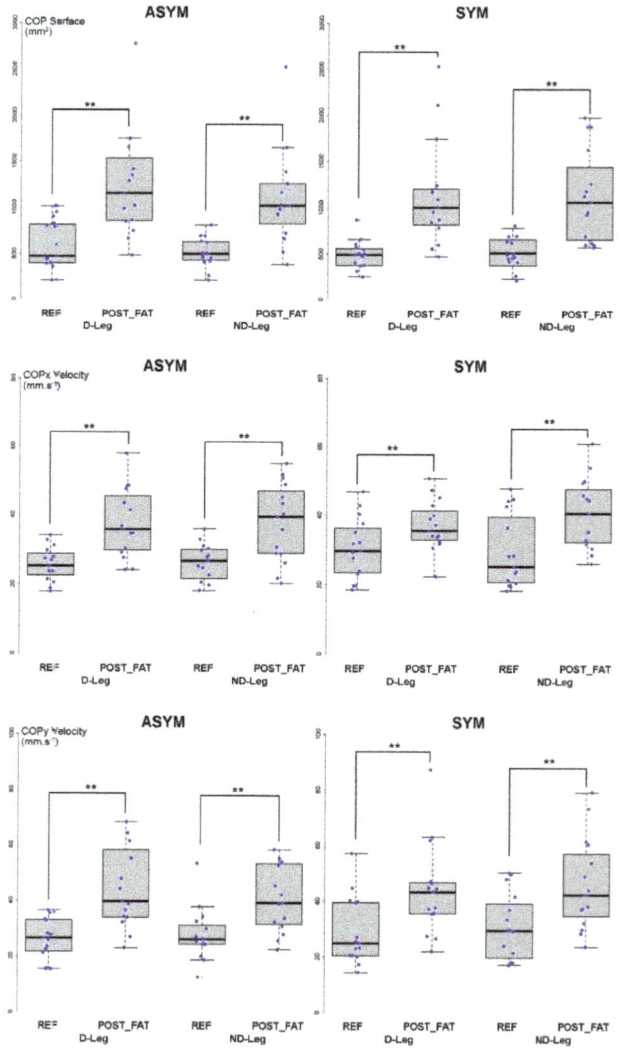

Figure 2. Boxplot representation with individual data points of postural parameter values on the D-Leg and ND-Leg in the ASYM and SYM groups in REF and POST_FAT conditions. Note, ** denotes a significant fatigue effect ($p < 0.01$) from the comparison between the REF and POST-FAT conditions. No significant differences were observed between the two legs.

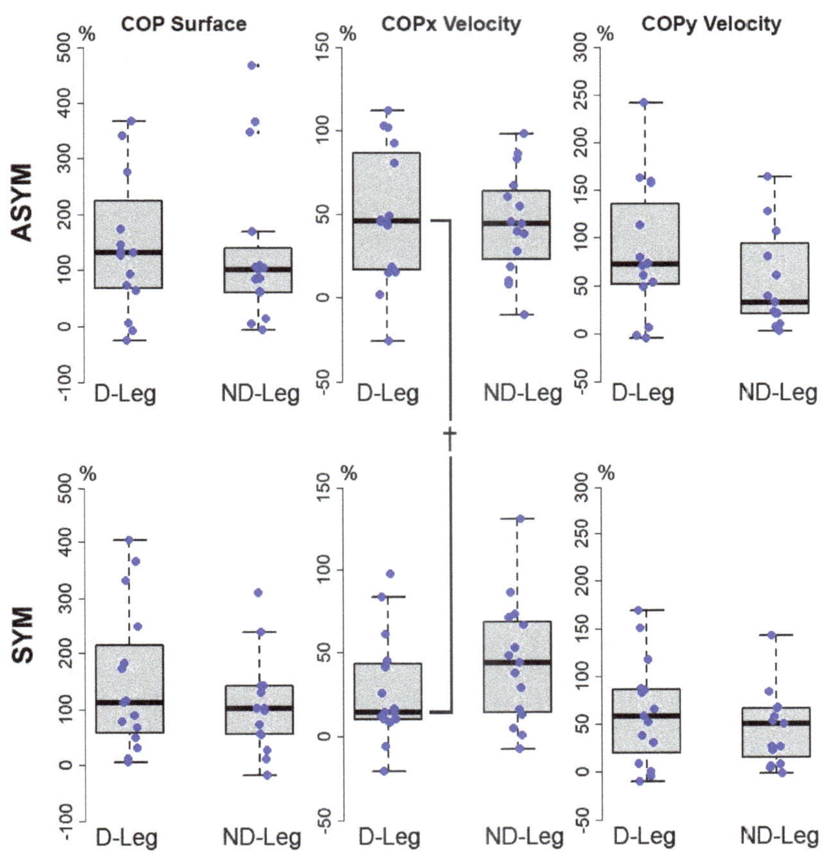

Figure 3. Boxplot representation with individual data points of the relative increases in the postural parameter values on the D-Leg and ND-Leg in the ASYM and SYM groups. Note, † indicates a tendency (p = 0.06) between the ASYM and SYM groups.

Fatigue effect: The COP surface, the COPx, and the COPy velocities increased more in the POST_FAT condition compared to the REF condition (Figure 2).

There were no differences between the groups with respect to the duration of the fatigue exercise protocol (ASYM, first period: 30.71 ± 22.56 min, second period: 3.09 ± 1.36 min; SYM, first period: 32.50 ± 28.47 min, second period: 3.94 ± 3.69 min (mean ± SD)).

Leg dominance effect: Statistical comparisons between the D-Leg and ND-Leg showed no difference in the two conditions for both the ASYM and SYM groups.

Group effect: No differences were initially observed between the ASYM and SYM groups under each condition. A strong tendency could be observed in the relative increase in COPx velocity on the D-Leg in the POST_FAT condition (Figure 3), which tended to be higher in the ASYM group than the SYM group (p = 0.06).

4. Discussion

The present pilot study was the first study to focus on differences in postural control between the D-Leg and the ND-Leg, following physiological states induced by fatiguing exercise, among ASYM and SYM athletes. The fatiguing exercise had a disturbing effect on monopedal postural control regardless of the leg used and the nature of sport practiced.

This disturbing effect related to fatigue tended to be more marked on the D-Leg in athletes practicing asymmetric sports than in athletes practicing symmetric sports.

The result of this study showed that a voluntary fatiguing exercise adversely affected postural control on both the D-Leg and the ND-Leg in both the SYM and ASYM groups. This supports previous studies that reported a deterioration of postural control regardless of the leg on which the subjects were assessed [16,22]. The fatigue exercise employed in the present study, which consisted of repeating squats until exhaustion, can be considered as a global exercise that solicits a large part of the body musculature [17]. With a mean total duration that exceeded 30 min in both the ASYM and SYM groups, fatigue induced by this type of exercise is likely to generate peripheral and central fatigue, which can alter sensory inputs (i.e., disturbance of proprioceptive myotatic information specifically related to the fatigue engendered at the level of the extensor muscles of the lower limb, and disturbance of the vestibular sensitivity specifically related to the organic and vestibular dehydration induced), their central integration (i.e., degradation of programming, command, and control of movement), and the motor output (i.e., decrease in muscle strength) of the postural function (for a review, see Paillard [17]). Further studies have shown that postural control is less affected by muscle fatigue on the ND-Leg than on the D-Leg, especially with athletes involved in ASYM activities, such as netball [18], basketball [5], or soccer [23]. Our results display concordant findings, since postural control on the D-Leg in the frontal plane tended to be impacted more in the presence of fatigue in the ASYM group than in the SYM group (COPx velocity, $p = 0.06$). Even if the sample sizes should have been larger, in order to expect clear significant results—since Cohen's index [24] only gave a small to medium effect size: d = 0.41— this result would be in line with the hypothesis formulated by Paillard [14], who postulated that the impact of limb dominance on monopedal postural control could be exacerbated by the specificity of motor experience (i.e., the practice of symmetric vs. asymmetric sports), and could be highlighted in the context of a negatively affected physiological condition, such as fatigue, and in the plane (frontal) in which monopedal posture is the most difficult to control. Indeed, there was initially no difference between the two groups on both the D-Leg and the ND-Leg (REF condition), and the tendency on the D-Leg between the SYM group and the ASYM group was only observed after the fatigue exercise in a lessened physiological condition (POST_FAT condition).

With asymmetric activities that require frequent phases of monopedal posture on the ND-Leg to perform technical movements with the D-Leg (e.g., passing and kicking in soccer), the ND-Leg can display better postural balance than the D-Leg [8,9,14]. In contrast, symmetric activities that use the two limbs similarly do not produce such an asymmetry of postural control [14]. This author inferred that particular motor tasks, regularly repeated, induce specific structural and functional adaptations at the central nervous system level, which generates durable modifications of motor and postural behaviors through a learning effect. As part of the study of cross-education, which attests that the motor output of the untrained limb (i.e., the contralateral limb) is improved after unilateral exercise training (i.e., the ipsilateral limb), it was reported that structural and/or functional differences at the cortical, subcortical and spinal levels were linked to the motor command between the dominant leg and the non-dominant leg [25,26]. For a given motor task, in its execution, the specialized limb (trained limb) would provide a better reference of motor information in the cortex, and would induce a better pattern of muscle activation (e.g., coordination of agonists and antagonists, synergist muscle activity, motor control) than the non-specialized limb (untrained limb in the execution of the considered motor task) [27]. These neurophysiological adaptations could enable postural control to be less affected on the ND-Leg (specialized leg) than on the D-Leg (non-specialized leg) in athletes practicing ASYM activities in adverse physiological conditions (fatigue), and in the direction (medio-lateral) that is the most difficult to control.

Thus, fatiguing exercise disturbed postural control on the D-Leg and ND-Leg in both the SYM and ASYM groups. This disturbance tended to be more marked on the D-Leg in athletes practicing asymmetric sports than in athletes practicing symmetric sports. However, this pilot study presents a certain limitation, and thus cannot provide a clear-cut answer to the question asked. At this end, future works dealing with the possible differences between the D-leg and the ND-leg, in terms of postural control, should be carried out with larger sample sizes, in order to possibly obtain clear and meaningful results. These future studies could also include additional experimental analyses, such as, for example, electromyographic measurements, in order to answer the question more precisely.

This article reports an innovative protocol including the nature of sport practiced and the physiological states in which the subjects were evaluated, in order to determine the possible differences between the D-Leg and the ND-Leg in terms of postural control. The D-Leg seems to be more sensitive than the ND-Leg in athletes practicing asymmetric sports than in athletes practicing symmetric sports. Therapists and trainers should be aware of the possible difference between the dominant leg and the non-dominant leg during their intervention protocol based on monopedal postural tasks, especially when the athletes are evaluated in a lessened physiological condition (e.g., immediately after strenuous exercise and/or a sports competition).

Author Contributions: Conceptualization, T.P.; methodology, T.P.; M.A.K. software, M.A.K., J.M. and F.N.; validation, M.A.K., F.N., J.M., N.M. and T.P.; formal analysis, M.A.K., J.M., F.N. and T.P.; investigation, M.A.K.; data curation, M.A.K.; writing M.A.K. and T.P. All authors have read and agreed to the published version of the manuscript.

Funding: This research received no external funding.

Institutional Review Board Statement: The study was conducted according to the guidelines of the Declaration of Helsinki.

Informed Consent Statement: Informed consent was obtained from all subjects involved in the study.

Acknowledgments: The authors would like to thank all the participants who took part in the present study.

Conflicts of Interest: The authors declare no conflict of interest.

References

1. Cheung, R.; Smith, A.; Wong, D. H: Q ratios and bilateral leg strength in college field and court sports players. *J. Hum. Kinet.* **2012**, *33*, 63–71. [CrossRef] [PubMed]
2. Hoffman, M.; Schrader, J.; Applegate, T.; Koceja, D. Unilateral postural control of the functionally dominant and nondominant extremities of healthy subjects. *J. Athl. Train.* **1998**, *33*, 319–322. [PubMed]
3. Zazulak, B.T.; Ponce, P.L.; Straub, S.J.; Medvecky, M.J.; Avedisian, L.; Hewett, T.E. Gender comparison of hip muscle activity during single-leg landing. *J. Orthop. Sports Phy. S Ther.* **2005**, *35*, 292–299. [CrossRef] [PubMed]
4. Paillard, T.; Noé, F. Does monopedal postural balance differ between the dominant leg and the non-dominant leg? A review. *Hum. Mov. Sci.* **2020**, *74*, 102686. [CrossRef] [PubMed]
5. Erkmen, N.; Suveren, S.; Göktepe, A. Effects of Exercise Continued Until Anaerobic Threshold on Balance Performance in Male Basketball Players. *J. Hum. Kinet.* **2012**, *33*, 73–79. [CrossRef]
6. Guillou, E.; Dupui, P.; Golomer, E. Dynamic balance sensory motor control and symmetrical or asymmetrical equilibrium training. *Clin. Neurophysiol.* **2007**, *118*, 317–324. [CrossRef]
7. Marchetti, P.H.; Orselli, M.I.V.; Martins, L.; Duarte, M. Effects of a full season on stabilometric Parameters of team handball elite athletes. *Motriz. Rev. Ed. Fis.* **2014**, *20*, 71–77. [CrossRef]
8. Ricotti, L.; Rigosa, J.; Niosi, A.; Menciassi, A. Analysis of balance, rapidity, force and reaction times of soccer players at different levels of competition. *PLoS ONE* **2013**, *8*, e77264. [CrossRef]
9. Gstöttner, M.; Neher, A.; Scholtz, A.; Millonig, M.; Lembert, S.; Raschner, C. Balance ability and muscle response of the preferred and nonpreferred leg in soccer players. *Motor. Control.* **2009**, *13*, 218–231. [CrossRef]
10. Huurnink, A.; Fransz, D.P.; Kingma, I.; Hupperets, M.D.; van Dieën, J.H. The effect of leg preference on postural stability in healthy athletes. *J. Biomech.* **2014**, *47*, 308–312. [CrossRef]
11. Matsuda, S.; Demura, S.; Demura, T. Examining differences between center of pressure sway in one-legged and two-legged stances for soccer players and typical adults. *Percept. Mot. Ski.* **2010**, *110*, 751–760. [CrossRef] [PubMed]
12. Matsuda, S.; Demura, S.; Uchiyama, M. Centre of pressure sway characteristics during static one-legged stance of athletes from different sports. *J. Sports Sci.* **2008**, *26*, 775–779. [CrossRef]

13. Sabin, M.J.; Ebersole, K.T.; Martindale, A.R.; Price, J.W.; Broglio, S.P. Balance performance in male and female collegiate basketball athletes: Influence of testing surface. *J. Strength. Cond. Res.* **2010**, *24*, 2073–2078. [CrossRef]
14. Paillard, T. Plasticity of the postural function to sport and/or motor experience. *Neurosci. Biobehav. Rev.* **2017**, *72*, 129–152. [CrossRef]
15. Barone, R.; Macaluso, F.; Traina, M.; Leonardi, V.; Farina, F.; Di Felice, V. Soccer players have a better standing balance in nondominant one-legged stance. *Open Access J. Sports Med.* **2011**, *2*, 1–6. [CrossRef]
16. Marchetti, P.H.; Orselli, M.I.; Duarte, M. The effects of uni-and bilateral fatigue on postural and power tasks. *J. Appl. Biomech.* **2013**, *29*, 44–48. [CrossRef] [PubMed]
17. Paillard, T. Effects of general and local fatigue on postural control: A review. *Neurosci. Biobehav. Rev.* **2012**, *36*, 162–176. [CrossRef]
18. Waterman, N.; Sole, G.; Hale, L. The effect of netball game on parameters of balance. *Phys. Ther. Sport* **2004**, *5*, 200–207. [CrossRef]
19. Cug, M.; Wikstrom, E.A. Learning effects associated with the least stable level of the biodex® stability system during dual and single limb stance. *J. Sport Sci. Med.* **2014**, *13*, 387–392.
20. Harkins, K.M.; Mattacola, C.G.; Uhl, T.L.; Malone, T.R.; Mccrory, J.L. Effects of 2 ankle fatigue models on the duration of postural stability dysfunction. *J. Athl. Train.* **2005**, *40*, 191–194. [PubMed]
21. Paillard, T.; Noé, F. Techniques and methods for testing the postural function in healthy and pathological subjects. *BioMed Res. Int.* **2015**, *2015*, 891390. [CrossRef] [PubMed]
22. Brito, J.; Fontes, I.; Ribeiro, F.; Raposo, A.; Krustrup, P.; Rebelo, A. Postural stability decreases in elite young soccer players after a competitive soccer match. *Phys. Ther. Sport* **2012**, *13*, 175–179. [CrossRef] [PubMed]
23. Arliani, G.G.; Almeida, G.P.; Dos Santos, C.V.; Venturini, A.M.; Astur Dda, C.; Cohen, M. The effects of exertion on the postural stability in young soccer players. *Acta Ortop. Bras.* **2013**, *21*, 155–158. [CrossRef] [PubMed]
24. Cohen, J. *Statistical Power Analysis for the Behavioral Sciences*, 2nd ed.; Á/L. Erbaum Press: Hillsdale, NJ, USA, 1988.
25. Carroll, T.J.; Herbert, R.D.; Munn, J.; Lee, M.; Gandevia, S.C. Contralateral effects of unilateral strength training: Evidence and possible mechanisms. *J. Appl. Physiol.* **2006**, *101*, 1514–1522. [CrossRef]
26. Frazer, A.K.; Pearce, A.J.; Howatson, G.; Thomas, K.; Goodall, S.; Kidgell, D.J. Determining the potential sites of neural adaptation to cross-education: Implications for the cross-education of muscle strength. *Eur. J. Appl. Physiol.* **2018**, *118*, 1751–1772. [CrossRef]
27. Farthing, J.P. Cross-education of strength depends on limb dominance: Implications for theory and application. *Exerc. Sport Sci. Rev.* **2009**, *37*, 179–187. [CrossRef]

symmetry

MDPI

Article

From Neural Command to Robotic Use: The Role of Symmetry/Asymmetry in Postural and Locomotor Activities

Mariève Blanchet [1], Pierre Guertin [2], Francine Pilon [3], Philippe Gorce [4,5] and François Prince [5,6,*]

1 Département des Sciences de L'activité Physique, Université du Québec à Montréal, Montréal, QC H3C 3P8, Canada; blanchet.marieve@uqam.ca
2 Département de Psychiatrie et Neurosciences, Faculté de Médecine, Université Laval, Québec, QC G1V 0A6, Canada; pierre-a.guertin@fmed.ulaval.ca
3 École de Kinésiologie et des Sciences de L'activité Physique, Université de Montréal, Montréal, QC H3C 3J7, Canada; francine.pilon@umontreal.ca
4 Université de Toulon, 83130 La Garde, France; philippe.gorce@univ-tln.fr
5 International Institute for Biomechanics and Surgical Ergonomics (INCISION), Université de Toulon, 83130 Toulon, France
6 Département de Chirurgie, Faculté de Médecine, Université de Montréal, Montréal, QC H3C 3J7, Canada
* Correspondence: francois.prince@umontreal.ca

Abstract: This article deepens a reflection on why and how symmetry/asymmetry affects the motor and postural behavior from the neural source, uterine development, child maturation, and how the notion of symmetry/asymmetry has been applied to walking robot design and control. The concepts of morphology and tensegrity are also presented to illustrate how the biological structures have been used in both sciences and arts. The development of the brain and the neuro-fascia-musculoskeletal system seems to be quite symmetric from the beginning of life through to complete maturity. The neural sources of movements (i.e., central pattern generators) are able to produce both symmetric or asymmetric responses to accommodate to environmental constraints and task requirements. Despite the fact that the human development is mainly symmetric, asymmetries already regulate neurological and physiological development. Laterality and sports training could affect natural musculoskeletal symmetry. The plasticity and flexibility of the nervous system allows the abilities to adapt and compensate for environmental constraints and musculoskeletal asymmetries in order to optimize the postural and movement control. For designing humanoid walking robots, symmetry approaches have been mainly used to reduce the complexity of the online calculation. Applications in neurological retraining and rehabilitation should also be considered.

Keywords: symmetry; asymmetry; human; development; locomotion; posture; walking robot

Citation: Blanchet, M.; Guertin, P.; Pilon, F.; Gorce, P.; Prince, F. From Neural Command to Robotic Use: The Role of Symmetry/Asymmetry in Postural and Locomotor Activities. *Symmetry* **2021**, *13*, 1773. https://doi.org/10.3390/sym13101773

Academic Editor: Thierry Paillard

Received: 8 July 2021
Accepted: 17 September 2021
Published: 24 September 2021

1. Introduction

In order to achieve locomotor activity with a high level of symmetry [1–4], we can ask ourselves whether the components of our musculoskeletal system absolutely need to be symmetrical and whether the neural control that activates the functionality of our muscles should also be symmetrical. Figure 1 presents the conceptual elements taken into consideration in how we approach the notion of symmetry/asymmetry in the control of posture and locomotion. In this article, we will introduce the concepts of tension imbalance, tensegrity, and dysmorphism to illustrate how the biological structures have been influenced and altered during growth and maturation, the development of laterality, as well as the influence of task requirements and the environmental constraints. This article will also deepen the reflection on why and how symmetry/asymmetry affects locomotor and postural behaviors. Finally, we will present how the neural source for locomotion and the early humanoid walking robot commands were assuming symmetry and how this command was adapted to more human-like environmental constraints and task-oriented requirements.

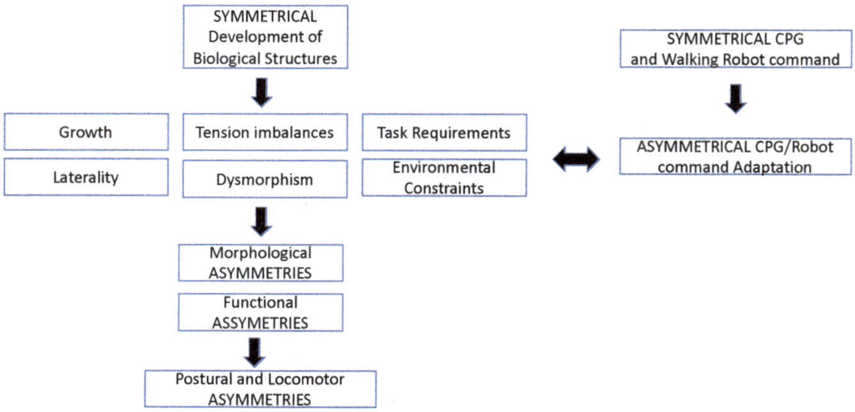

Figure 1. Conceptual elements taken into consideration in the symmetry/asymmetry approach in the control of posture and locomotion. CPG: central pattern generator.

2. Symmetry/Asymmetry in Human as Seen from Arts and Functional Anatomy

What is beautiful is symmetrical. From an aesthetic point of view, this is undoubtedly true. In general, symmetry is seen as attractive. In the classical Greek period, the Greeks possessed a sense of beauty. Indeed, Greek sculptures from the classical period such as the Lancelotti Discobolus (National Museum Rome) clearly demonstrate this sense of beauty. Greek statues and Renaissance sculptures, not to mention Leonardo da Vinci's Vitruvian Man, represent perfectly symmetrical human forms. It has long been the human morphological standard, and perhaps still is. During the Renaissance, Michelangelo's sculpture of David captured balance and harmony of forms and brought to the fore the notion of ideal form. Even to this day, bipedal postural analysis is based on this right/left symmetry of form as well as straightness and divergence of lines. Additionally, according to the principle of tensegrity (contraction of the words tension and integrity), a structure is stabilized by a continuous tension applied to discontinuous elements in compression. The architectural principle called "tensegrity" and put forward by Buckminster Fuller in the 1960s was inspired by a sculpture made by sculptor Kenneth Snelson in 1948 (the X-Piece). This principle is opposed to the traditional principle of man-made structures involving continuous gravitational compression. This implies that in weightlessness, these structures can lose their shape while the tensegrity structures retain it. Inter-limb symmetry/asymmetry may occur as a function of motor experience (e.g., high versus low), the nature of movements (e.g., specialized versus non-specialized), the environmental context (e.g., easy vs. difficult motor tasks), individual/intrinsic factors (e.g., afferences, hemispheric laterality, and motor output), and the limb dominance effect. However, on the one hand, the finer details of motor and postural symmetry/asymmetry have not yet been fully identified in terms of information perception, central integration, and movement command and control. On the other hand, the neural mechanisms involved are also not fully understood at the different neurological levels (peripheral, spinal, subcortical, and cortical). Therefore, exploratory research is needed in order to understand symmetry/asymmetry in terms of human movement and posture.

The perfect shape is desired by nature, and as long as the tensor elements, which are the skin, muscles, and fascia, have an optimal length, the symmetry of the shape will be respected. But what is it from a functional point of view? Why is right/left inter-limb symmetry so particular to human movement and posture?

First of all, where did the idea of symmetry come from? The notion of left/right symmetry has been a very well-defined concept, especially in the case of the human body, since the mid-1900s [5]. The common meaning given to bilateral symmetry is based on the notion of proportionality, balance, and concordance between the parts in order to form a whole. Looking at a given population, it is difficult to conceive that there is a single human morphology, a theoretical ideal shape that expresses left/right symmetry. However, in 1949, Françoise Mézières entertains the notion that there is a human morphology known as "normal" and symmetrical and all forms deviating from this human morphology constitute a set of characteristics allowing identifying dysmorphisms called biotypes [6]. In our classical language, there is no so-called "normal" morphology, but only various morphotypes such as ectomorph, endomorph, and mesomorph used to define persons according to their genetics. We can therefore consider that the human morphological phenotype is the expression of all genes. If human morphological symmetry exists, where does it come from? How is it defined?

"The molecules that make up cells and cells that make up tissues are continually renewing and the maintenance of the integrity of this behavior is living. This behavior is a manifestation of structure and structural stability and resulted in the establishment of spatial relationships that balance the individually destabilized structural elements" [7]. Undeniably, at the level of its neuro-fascia-musculoskeletal system, the human body corresponds to the definition of symmetry, namely, the right side is the identical reflection of the left side. We find exactly the same bone structures and the same myofascial structures on both sides of the body. In the frontal plane, the median axis of the body divides the body in two and highlights this symmetry whether the gaze is projected on the anterior or posterior part of the body. Symmetry brings a certain stability. Symmetrical structures therefore make it possible to distribute forces equally throughout the neuro-musculoskeletal system. The anatomical continuities between the different muscles allow reciprocal feedback to take place through multiple mechanical and nervous pathways, activation time, intensity, duration, and release of tissue deformation, the latter being precisely controlled by a variety of sensory inputs such as proprioceptors located in connective tissue.

The introduction of new concepts such as those related to the fascia system caused the musculoskeletal duality to become obsolete and replaced by the notion of the mesokinetic system binding bones, muscles, and connective tissues into a symmetrical and functional unit [8,9]. The mesokinetic system is a unifying structural whole. In this unifying whole, we find junctions (i.e., joints) of remarkable precision that intertwine with each other in order to provide a dynamic of free, flexible movement [10,11]. Because the entire system works as a whole, the functioning of the body can be revealed, regardless of the situation or position. Thus, the entire system from the cytoskeleton to the mesokinetic system plays a unifying role in the common goal of structural integrity and movement [12]. However, observation of human morphology reveals a more asymmetric aspect.

Human beings are complex organisms and at the same time have a relatively simple geometry constituting a complete functional unit. At the heart of this functional unity lies the very notion of symmetry. What the left/right morphological asymmetry represents in typical human adults is dysmorphisms and tension imbalances in the myofascial elements, not structural asymmetry. The tension imbalances in the myofascial elements could be explained, in part, by the directional asymmetry of internal organs. According to Klingenberg [13], the internal organs of the human body are organized according to a directional asymmetry, that is, the traits develop differently on the left and right side of the body, for example, the lungs, which have three lobes on the right side and two lobes on the left side. Moreover, the developmental rhythm of neuro-fascia-musculoskeletal system could also impose a certain form of asymmetry, but dysmorphism and tension imbalances should be considered the main sources of morphological and functional asymmetries. In this section, we have shown that structurally, the human body is developed with a symmetrical pattern. Morphological asymmetries are the results of myofascial imbalances and produce functional asymmetries in posture and locomotion.

Symmetry and Asymmetry in Humans: From Uterine Development to Adult Life

From the beginning of intrauterine life, symmetry/asymmetry regulates development. The third trimester promotes the flexed position when the infant is crowded by the uterine environment and experiences rapid brain growth, mediating flexion (arms and legs bent and trunk tucked forward) [14]. Despite this temporal frame corresponding to non–goal-directed fetal motility [15], positioning in physiological flexion (flexion of the shoulders, hips, and knees, scapular protraction, and posterior pelvic tilt) is the ideal position of the newborn, as it promotes proper symmetrical joint alignment, supports neuromuscular development, and promotes self-soothing and behavioral organization [16,17]. Interestingly, the innate genetic instructions indicated an asymmetric anterior/posterior development where the antigravity muscles responsible for antagonist movements (extensor posterior postural adjustments) mature earlier compared to the flexors [15]. Indeed, many studies showed that tibialis anterior (TA) (ankle flexor muscles) plays a different role in the postural control of children compared to adults [18,19]. Berger's group [20] explained this ontogenetic difference by a more central regulation of flexor muscle activity compared to the extensor, which has an effective circuitry in the lower levels. Furthermore, the pathways that innervate TA muscles mature later than gastrocnemius (posterior muscle) pathways despite their similar distal localization with respect to the ankle [19] and create asymmetries in motor and postural behaviors.

During growth, a gradual symmetry in the body, organ, and tissue (lengths, areas, and volumes) can be observed. The development of the human brain is also a long-lasting process, which is mirrored by a multitude of developmental changes such as in motor behavior [15]. In fact, these neurological mechanisms evolve over time in a non-linear way in which we can observe a sudden rather than a gradual change with age [21]. The functional symmetrical topography of the brain is primarily driven by genetic instructions, the starting point for epigenetic cascades that allow abundant interactions with the environment and activity-dependent processes [22,23]. The interaction is bidirectional where experience affects gene expression and genes affect how the environment is experienced [23]. The environment and activity-dependent processes shape the brain, and a certain asymmetry could appear especially in the cortical homunculus mapping (for example, in the musician). The abundance of cerebral connectivity is the neural basis of human behavioral variability, i.e., the ability to select, from a large repertoire of behavioral solutions, the one most appropriate for a specific situation [15]. This flexible and adaptative neurological capacity allows the possibility to adapt their movement responses to the symmetric/asymmetric biomechanical demands from the task requirements and the environmental constraints. The period when major and rapid postural symmetric/asymmetric of growth changes occur corresponds to the time when the cerebral plasticity is increased (i.e., before adult age). Indeed, childhood and adolescence are sensitive developmental periods associated with an increasing sensorimotor experience leading to a different effect on motor behavior [24,25].

The non-monotonic pattern that dictates the rhythm of motor development of several parameters has been reported in studies assessing reactive postural adjustments [26], postural control adjustments during self-initiated unloading [27], goal-directed arm movements [28], stability limits [29], and quiet standing tasks [21,30–33]. Increasing evidence indicates that this period corresponds with a critical transition period for maturation (around 6 or 7 years of age).

One important hypothesis that has been proposed to explain these sudden changes in movement and postural control during the transition period is associated with nervous system adaptations in which the effectiveness of the processing and integration of multimodal sensory information increase and evolve from an en bloc strategy (also named the ballistic strategy) from 0 to 5 years of age toward a sensory strategy that is mastered over 8 years [32,34]. However, it is possible that the transition period was a necessary sensory recalibration period after rapid development of the body segments in order to update the internal model (body image).

In fact, the growth of the different segments is not uniform and symmetrical. Surprisingly, a certain genetic asymmetry regulates the course of the lower limb. For example, the analysis of 354 unaffected hip–knee–ankle angles with anteroposterior full-length standing radiographs revealed that participants aged from 1 to 2 years old were naturally in varus (+3.6°) during the emergence of locomotor functions. However, the following year (2 to 3 years old) undergoes a drastic change in the hip–knee–ankle angles from 5.1° in the opposite direction in order to reach a valgus posture (means, −2.5°) [35]. This period corresponds to the development of dynamic postural control mechanisms, which allows controlling bipedal body posture during displacement and during active movements [36]. Postural control is intimately linked to motor control: dynamic motor actions cannot be performed without first stabilizing body posture [37]. In order to compensate for this asymmetric postural development and inexperience, children select the en bloc strategy that allowed the possibility to limit the degrees of freedom and facilitate the direction-specific postural muscles recruitment [15]. The en bloc strategy is dominant often between the ages of 9 months and 2.5 years old and is largely used until the transition period [19,34], corresponding with the drastic change in the lower limbs angle configuration. After the transition period, around the age of 8 years old, the lower limbs angle reaches the one of adults (i.e., varus posture of +11.2°) [35], and both populations use the sensory strategy [34].

During skeletal development, bones increase in size and mineral mass while their morphology adapts according to genetics and to mechanical constraints from the task demands and environmental factors [38].

Similarly, when researchers compared the bone mineral content and the bone density at a stressed bone site (the dominant arm in a tennis or squash player) with little or no bone solicited from the site of their non-dominant arm, the results show differences ranging from 10% to 15% after only a few years of practice [39].

It is no wonder that the development of laterality (neurological factors) can influence the development of morphological asymmetries.

Laterality is a complex concept. It is expressed in predominantly manual, ocular, pedal, and auditory preferences, differences in sensorimotor performance between preferred and non-preferred effectors, and directional tendencies. It is one of the expressions of functional hemispherical asymmetries [40] that defines functional superiority on one side. Genetically determined at birth, the majority of people [41] have a match between the hand used to write, the foot used to kick the ball, and the eye used to look through a telescope. A typical young child with a manual predominance will choose the writing hand in a spontaneous and natural way. It emerges around the age of 3 to 3.5 years [42] and continues to refine itself until the beginning of adolescence (laterality represented and projected in the absence of the object or of the person). Laterality is therefore part of the evolution of the bone growth and of gross and fine motor skills asymmetry.

In light of these postulates, the body representation (internal model) is possibly the most important link between symmetric/asymmetric morphological changes and their influence on movement and postural control. It assumes the existence of an internal representation of the "geometry of the body", the ground reaction forces, and its orientation relative to the vertical [43]. The early perception–action coupling is a fundamental process that allows the correspondence between the perception of an action, its sensorimotor representation, and its realization [44]. This body representation develops during childhood through the regular and varied interactions of the senses, especially with proprioceptive information [44,45]. Overall, this highlights the importance of regular and varied experience for all populations, especially in children, in order to continuously update the body representation and reinforces the need to avoid early sports specialization.

Body segments, organs, and tissues develop in a symmetrical pattern from uterine to early childhood periods. Then, laterality and motor skills are developed under the influence of both environmental constraints and task requirements. Evolution of the early postural control patterns "en bloc" is progressively modified to a more adaptative and mature response.

3. A Central System for Locomotor Rhythm and Pattern Generation: Control of Symmetrical/Asymmetrical Activities by the Spinal Cord

More than a century ago, Graham Brown provided compelling evidence that locomotion was essentially controlled by a neuronal network located in the spinal cord [46,47]. In anesthetized cats, rabbits, or guinea pigs, he showed spontaneously occurring hindlimb stepping movements after a complete transection (Tx) of the spinal cord at the thoracic level. Given that (1) doses of anesthetic used by Graham Brown were known to abolish selectively proprioceptive and exteroceptive reflexes and (2) descending commands from the brain after a Tx could no longer exert control over hindlimb muscle contraction, he proposed the existence of a spinal command center located in lumbar segments, called the 'half-center', that would be responsible for locomotor rhythm and pattern generation in the lower limbs. He imagined the network to be composed of two groups of neurons, reciprocally connected and mutually inhibiting each other in such a way that activity in the first group (e.g., extensor half-center) would activate extensor muscles while inhibiting the reciprocal group of neurons (flexor half-center) for the concomitant relaxation of flexor muscles and execution of the stance phase. After a period of 'depression' of the extensor half-center due to fatigue (due to adaptation or post-inhibitory rebound), the second group of neurons (flexor half-center) would take over for the next phase of activity—e.g., the contraction of flexors, relaxation of extensors, and execution of the swing phase.

In the 1970s, the existence of such a central command center, thereafter called the Central Pattern Generator (CPG) for locomotion, was clearly demonstrated experimentally by Grillner and Zangger using completely deafferented spinal Tx animals [48,49]. In the meantime, another group of Swedish researchers obtained the first electrophysiological evidence of its existence in lumbar segments of the spinal cord (lamina VII) using intracellular recording techniques, L-DOPA injection, and flexion reflex afferent stimulation [50,51]. Still today, a plethora of studies are being conducted to identify further CPG elements and characteristics. Based on some of them, it is now generally accepted that the CPG is composed of genetically identified cells such as the HB9, V0, V1, V2, and Shox2 interneurons (for left—right coordination or rhythm and speed control), intrinsic cellular properties such as endogenous bursting neurons and Ih current (for pacemaker-like generation), specific pharmacological properties such as 5-HT1 and D1 receptors (for CPG activation), and complex network connections that support synaptic interactions as those proposed in the ring model, flexor burst model, or two-level organization model for distinct and selective rhythm and pattern adaptation [52].

In normal conditions, basic locomotor gaits such as straightforward walking at low speed are generally considered to be more or less symmetrically organized—that is, with a steady rhythm, pattern, and timing of muscle activity. For instance, at the ankle level, the medial gastrocnemius (extensor) will be typically contracted throughout stance and relaxed during swing with a rather strict out-of-phase relationship with its direct antagonist, the tibialis anterior. At other joints of the limb, comparable alternating out-of-phase relationships will also be found between agonists (extensors) and antagonists (flexors) unilaterally as well as between homonymous muscles bilaterally (left and right biceps femoris) during bipedal walking [53]. This said, multiple symmetrical patterns and gaits exist given that a wide variety of strategies can be used by animals, including humans, to move from A to B by swimming, flying, using bipedal or quadrupedal walking, running, or galloping. Yet, clear evidence shows that, among all vertebrate species, all gaits and forms of stereotyped rhythmic motor behaviors are similarly controlled by central centers such as the CPG in association with other sets of neurons located in the mesencephalic locomotor region (MLR) of the brainstem and elsewhere in the nervous system [54].

Otherwise, many conditions also exist for which asymmetrical muscle contraction can be found. Depending on species, gaits, goals, and/or imposed demands such as pathologies, amputation, overloading, or directional changes, different patterns of muscle activity have been reported. For instance, rather abnormal and more or less asymmetrical patterns were found in people diagnosed with the Uner Tan Syndrome expressing quadrupedal walking [55]. In four-legged animals, patterns of muscle activity often differ considerably between slow (walking) versus fast locomotion (trotting, galloping), suggesting, in turn, the existence of a speed- or task-dependent reorganization of the CPG. For directional changes, asymmetrical or atypical muscle contraction is found, for example, to turn left during swimming; specific stimulation of some reticular formation nuclei, the Middle Rhombencephalic Reticular Nucleus, generally elicits, within a few milliseconds, a C-shape contraction of the entire body on the side of its new trajectory, momentarily replacing and resetting the regular left–right rhythmic pattern of axial muscle activity, which has led to the suggestion of a key role for this brainstem area in directional changes during locomotion [56]. Stimulation of other brainstem areas and nuclei was also found to trigger locomotor adaptations such as speed increase (Posterior Rhombencephalic Reticular Nucleus stimulation) or highly specific directional change and asymmetrical pattern (Anterior Rhombencephalic Reticular Nucleus stimulation for contralateral turns) [57]. Control over both initiation and speed adaptation has also been shown following stimulation of the MLR. Russians in the 1960s showed in decerebrated cats that weak stimulation of the MLR tonically, at the mesopontine junction of the dorsal reticular formation, could elicit walking in decerebrate cats, whereas stronger stimulation led to greater speeds and hence to gait alterations such as galloping instead of walking [58,59].

That great flexibility in pattern, speed, and gait enabled by CPG interactions with other structures including brainstem nuclei is not limited to the CNS. For instance, speeds can also be partially altered by peripheral-input-induced CPG mediated actions. One of the most relevant evidence has come from Forssberg and colleagues in the 1980s using Tx kittens walking on a two-belt treadmill—they showed that increasing speeds of only one belt did not prevent the other leg from walking 'normally' at lower speeds on the other belt [60]. Comparable observations made recently in adult cats suggest that adaptations of that nature, probably involving joint afferent inputs, remain possible in mature and chronically injured animals [61]. Other experiments with Tx cats performed by Forssberg [62] also showed that one leg perturbed after hurting an obstacle can express a bilaterally coordinated hyperflexion that brings the foot above and over the obstacle in order to maintain successful walking. Other receptor systems such as the proprioceptors have also been shown to play a pivotal role in CPG adaptation and asymmetrical control. When stimulated electrically during locomotion, muscle spindles (Ia afferents) and Golgi tendon organs (Ib afferents) were shown to enable extensive coordinated corrective responses expressed throughout the legs bilaterally—a form of temporary cycle-to-cycle asymmetrical adaptation in response to a sudden external disturbance (e.g., hole or overload). Only during locomotion (i.e., not at rest), group I afferents (Ia and Ib) from ankle extensors [63] or group II afferents from flexors [64], when stimulated briefly (100 ms), lead to CPG-mediated responses by replacing the correspondent classical reflex actions that promote the activity of extensors while inhibiting flexors throughout the entire limb in decerebrate cats.

All in all, the findings described above about locomotor-dependent responses (speed increase, gait alteration, directional change, extension enhancement, step cycle resetting, obstacle avoidance, hole, etc.) detected or stimulated suddenly by brief activation of specific peripheral receptor systems (cutaneous, joint, muscle Ia or Ib afferents) and/or central supraspinal structures (e.g., reticular formation, MLR, visual system, etc.) constitute examples that provide compelling evidence that the CPG is endogenously all set and prepared for a wide variety of symmetrical demands and of how these responses can be adapted with asymmetrical corrections under various environmental conditions and task requirement circumstances.

4. Symmetry/Asymmetry from Robotic Point of View

Researchers studying human gait have approached its quantitative evaluation through various parameters: stride speed, stride length, step information (length, width, angle and time), joint angles, muscle strength, etc.). Regardless of the field of application (rehabilitation, sport, or robotics (humanoid or walking robot)) the factors (indices) that allow its qualification are static or dynamic balance, stability, and symmetry/asymmetry, either to find a so-called healthy (or "normal") gait, maximize performance, or simply reproduce human walking with a humanoid robot. They are characterized by the trajectory of the joints, left/right symmetry, center of gravity (CoG), center of pressure (CoP), ZMP, etc.; international research has provided many works that lead to two different approaches (or models).

The first is the oldest and most used. It consists of describing walking as a continuous sequence of cyclic articular rotations of the limbs and trunk [65,66]. This concept is generally applied, with left/right symmetry as a prerequisite, to robotic systems (to mimic and reproduce human walking movement). In this context, the stability index is always estimated by monitoring whether the expected performance remains stable. The imitation of walking is, however, limited by the complexity of the body structure and its controls.

The second is less widespread and can be defined as the "forward translation" of the body system through "total locomotion". This movement results from the interaction between gravity, inertia, joint rotations, and the cyclic contraction of many muscles [67,68]. In this context, gait balance and symmetry are considered essential to maintaining gait translation performed by individual parts of the body. However, there is not yet a quantitative index to describe balance and symmetry, although many studies have been performed in this area.

Regardless of the approach chosen, the problem of asymmetry has often been studied and/or observed in people with neuromuscular pathology or alteration [65]. However, studies suggest that able-bodied people also sometimes exhibit asymmetric behaviors [2,69]. Understanding when and why this phenomenon occurs is important for gait research, where symmetry is typically assumed in order to simplify data collection and analysis. Many methods exist to quantify asymmetric movement between the right and left legs, using variables such as stride length [70–72], range of motion of the joints [73–75], velocity profiles [76] and ground reaction forces (GRF) [77–79], electromyographic profiles [80,81], limb forces and moments [82–84], or the oscillating center of mass [85,86]. However, the underlying causes are still the subject of debate. The functional asymmetry of gait hypothesis for able-bodied people suggests that each leg performs different roles, such as vertical support, medio-lateral (ML) control, and/or anteroposterior (AP) propulsion [2]. Differences between the roles of the legs have been observed in trials of brisk walking [87,88], suggesting that difficult locomotor tasks require asymmetric strategies. This is illustrated by asymmetry ratios in athletic walking [69] and running and cycling [89], which have been attributed to irregularities in the ground, footwear, and conditioning on the trails' curves. The most common explanation for functional asymmetry is leg dominance, but conflicting reports exist for this theory [71].

In the context of robotics (walking robot and humanoid), symmetry essentially constitutes a strong hypothesis for reducing the number of parameters that characterize and define walking. This is explained by the fact that a humanoid walking model has a high degree of redundancy that can be solved or bypassed by fixing the values of certain parameters, by resorting to optimization, or by adding constraint equations. As a result, being inspired by human walking can constitute a means of lowering the degree of redundancy in a simple way, allowing the humanoid to acquire a more natural behavior, and he can have characteristics of lower consumption of energy on unstructured soil [62,90].

In this context, the management of the balance associated with symmetry is a recurring and unavoidable problem for the generation of bioinspired walking for humanoids. The latter can be static or dynamic. In static walking, the projection of the CoM in the horizontal plane must be permanently inside the support polygon (convex envelope including the points of contact between the feet and the ground). In the case of dynamic walking, dynamic stability is obtained by using various criteria, including, among the most used, ZMP (zero moment point) and CWS (contact wrench sum). The ZMP is the point of the ground where the resultant of the reaction of the ground produces a zero moment along the anteroposterior and transverse axes [91–94]. This widely used criterion in the dynamic balance control of bipedal robots involves keeping the ZMP within the support polygon to prevent the foot from tipping over. Li and collaborators proposed a method that changes the position of the trunk when the ZMP deviates from the trajectory [95]. Other authors solve the problem of balance by modifying the position of the upper body [96–98] or the orientation of the trunk [99] or the waist position [100]. We can also note the use of the center of mass (CoM) to dynamically control a biped. The principle consists of controlling/regulating the acceleration of the CoM and ensuring a good distribution of the interaction forces between the segments and the trunk [101–104]. The CP (crossing point) refers to a virtual point located between the hip line and the ankle with each line to the left and right, respectively. The CWS was proposed [105] in the context of legged robots; it is based on the sum of the forces applied to the robot's CoM: if the sum of the forces of gravity and inertia applied to the CoM is inside the polyhedral convex cone of the contact forces between the robot's foot and the environment, then the balance is guaranteed. He extended and used this criterion on a humanoid on flat ground, to climb stairs, or on uneven ground [105]. We can also note the crossing point (CP) proposed by Kim and collaborators [106] and the foot rotation indicator (FRI) proposed by Goswani and their colleagues as another index used [107].

Another aspect to consider is that body mechanics are responsible for functional asymmetry. Simulation work has shown that the momentum and gravity are sufficient to propel the walking movement on a low slope [108,109]. These passive models reflect certain characteristics of human walking, such as ballistic movement during the oscillation phase [110] and energy efficiency on low slopes [111], and therefore act as simple substitution models for the study of bipedal mechanics. Although the dynamic equations of motion can give a stable solution corresponding to a symmetrical gait, small changes in the model parameters can result in qualitatively different behaviors at a bifurcation point, after which a new asymmetric (stable) solution emerges from the symmetric solution (unstable). The symmetrical mechanics of these walkers admit two families of solutions, one symmetrical and the other asymmetrical. However, the functions of these asymmetries have not been studied, and the period doubling phenomenon has not been shown to extend to more realistic 3D models that walk on flat ground.

On balance, symmetry/asymmetry serves to simplify and reduce the complexity of biological reality or a model, making it a powerful tool in robotic applications or related to computers and analytical modeling. Philosophically, does biological symmetry really exist? From this point of view, then one can question of the relevance of the models with a perfect symmetry—are they false for all that?

5. Conclusions

In conclusion, the development of the brain and the neuro-fascia-musculoskeletal system seem to be quite symmetric from the beginning of life through to complete maturity. The neural sources of movements, i.e., CPGs, are able to produce both symmetric or asymmetric responses to accommodate to environmental demands and task constraints.

Although the human development is mainly symmetric, asymmetries already regulate neurological and physiological development. The laterality and regular sports training could affect the natural musculoskeletal symmetry. The plasticity and flexibility of the nervous system allow the abilities to adapt and compensate for environmental constraints and musculoskeletal asymmetries in order to optimize the postural and locomotor control. For designing humanoid walking robots, symmetry approaches have been mainly used to reduce the complexity of the online calculation. With the improvement of computer power capacity progress, asymmetrical body models might be added in future walking robot developments. Applications in neurological retraining and rehabilitation should also be considered.

Author Contributions: Conceptualization: F.P. (François Prince); Methodology: F.P. (François Prince); Writing—original draft preparation, M.B., P.G. (Pierre Guertin), F.P. (Francine Pilon), P.G. (Philippe Gorce) and F.P. (François Prince); Writing—review and editing, F.P. (François Prince). All authors have read and agreed to the published version of the manuscript.

Funding: This research received no external funding. M.B. received a Junior 1 career scholarship from FRSQ.

Data Availability Statement: This paper was based on previous work published. No new data was presented.

Conflicts of Interest: The authors declare no conflict of interest.

References

1. Lindemann, U. Spatiotemporal gait analysis of older persons in clinical practice and research: Which parameters are relevant? *Z. Gerontol. Geriatr.* **2020**, *53*, 171–178. [CrossRef]
2. Sadeghi, H.; Allard, P.; Prince, F.; Labelle, H. Symmetry and limb dominance in able-bodied gait: A review. *Gait Posture* **2000**, *12*, 34–45. [CrossRef]
3. Sadeghi, H.; Prince, F.; Zabjek, K.; Labelle, H. Simultaneous, bilateral, and three-dimensional gait analysis of elderly people without impairments. *Am. J. Phys. Med. Rehabil.* **2004**, *83*, 112–123. [CrossRef]
4. Tucker, C.; Hanley, B. Increases in speed do not change gait symmetry or variability in world-class race walkers. *J. Sports Sci.* **2020**, *38*, 2758–2764. [CrossRef] [PubMed]
5. Weyl, H. *Symmetry*; Princeton University Press: Princeton, NJ, USA, 1952.thirty five years of its life. *Int. J. Hum. Robot.* **2004**, *1*, 157–173.
6. Geismar, S. *Mézières, Une Méthode, Une Femme. Le Dos Réinventé*; Josette Lyon: France, 1993.
7. Ingber, D.E. Cellular tensegrity: Defining new rules of biological design that govern the cytoskeleton. *J. Cell Sci.* **1993**, *104*, 613–627. [CrossRef]
8. Scarr, G. *Biotensegrity: The Structural Basis of Life*; Handspring Publishing Limited: Pencarirland, UK, 2014.
9. Scarr, G. Biotensegrity: What is the big deal? *J. Bodyw. Mov. Ther.* **2020**, *24*, 134–137. [CrossRef] [PubMed]
10. Myers, T.W. *Anatomy Trains: Myofascial Meridians for Manual and Movement Therapists*, 4th ed.; Elsevier: Amsterdam, The Netherlands, 2020.
11. Stecco, C. Why are there so many discussions about the nomenclature of fasciae? *J. Bodyw. Mov. Ther.* **2014**, *18*, 441–442. [CrossRef] [PubMed]
12. Stecco, C.; Gagey, O.; Belloni, A.; Pozzuoli, A.; Porzionato, A.; Macchi, V.; Aldegheri, R.; De Caro, R.; Delmas, V. Anatomy of the deep fascia of the upper limb. Second part: Study of innervation. Morphologie. *Bull. Assoc. Anat.* **2007**, *91*, 38–43. [CrossRef] [PubMed]
13. Klingenberg, C.P. Analyzing Fluctuating Asymmetry with Geometric Morphometrics: Concepts, Methods, and Applications. *Symmetry* **2015**, *7*, 843–934. [CrossRef]
14. Waitzman, K.A. The importance of positioning the near-term infant for sleep, play, and development. *Newborn Infant Nurs. Rev.* **2007**, *7*, 76–81. [CrossRef]
15. Hadders-Algra, M. Variation and Variability: Key Words in Human Motor Development. *Phys. Ther.* **2010**, *90*, 1823–1837. [CrossRef] [PubMed]
16. Aucott, S.; Donohue, P.K.; Atkins, E.; Allen, M.C. Neurodevelopmental care in the NICU. *Mental Retard. Dev. Disabilities Res. Rev.* **2002**, *8*, 298–308. [CrossRef] [PubMed]
17. Walicka-Cupry, K.; Szeliga, E.; Guzik, A.; Mrozkowiak, M.; Niewczas, M.; Ostrowski, P.; Tabaczek-Bejster, I. Evaluation of anterior-posterior spine curvatures and incidence of sagittal defects in children and adolescents practicing traditional karate. *BioMed Res. Int.* **2019**, *2019*, 9.
18. Blanchet, M.; Marchand, D.; Cadoret, G. Postural control adjustments during progressive inclination of the support surface in children. *Med. Eng. Phys.* **2012**, *34*, 1019–1023. [CrossRef]
19. Hadders-Algra, M.; Brogren, E.; Forssberg, H. Postural adjustments during sitting at preschool age: Presence of a transient toddling phase. *Dev. Med. Child Neurol.* **1998**, *40*, 436–447. [CrossRef] [PubMed]
20. Berger, W.; Discher, M.; Trippel, M.; Ibrahim, K.I.; Dietz, V. Developmental aspects of stance regulation, compensation and adaptation. *Exp. Brain Res.* **1992**, *90*, 610–619. [CrossRef]
21. Kirshenbaum, N.; Riach, C.; Starkes, J. Non-linear development of postural control and strategy use in young children: A longitudinal study. *Exp. Brain Res.* **2001**, *140*, 420–431. [CrossRef]
22. Changeux, J.P. Variation and selection in neural function. *Trends Neurosci.* **1997**, *20*, 291–293.
23. Diamond, A. The interplay of biology and the environment broadly defined. *Dev. Psychol.* **2009**, *45*, 1–8. [CrossRef]
24. Ismail, F.Y.; Fatemi, A.; Johnston, M.V. Cerebral plasticity: Windows of opportunity in the developing brain. *Eur. J. Paediatr. Neurol.* **2017**, *21*, 23–48. [CrossRef]

25. Paillard, T. Plasticity of the postural function to sport and/or motor experience. *Neurosci. Biobehav. Rev.* **2017**, *72*, 129–152. [CrossRef]
26. Shumway-Cook, A.; Woollacott, M.H. The growth of stability: Postural control from a development perspective. *J. Mot. Behav.* **1985**, *17*, 131–147. [CrossRef] [PubMed]
27. Hay, L.; Redon, C. Feedforward versus feedback control in children and adults subjected to a postural disturbance. *Exp. Brain Res.* **1999**, *125*, 153–162. [CrossRef] [PubMed]
28. Sigmundsson, H.; Whiting, H.T.; Loftesnes, J.M. Development of proprioceptive sensitivity. *Exp. Brain Res.* **2000**, *135*, 348–352. [CrossRef] [PubMed]
29. Riach, C.L.; Starkes, J.L. Stability limits of quiet standing postural control in children and adults. *Gait Posture* **1993**, *1*, 105–111. [CrossRef]
30. Figura, F.; Cama, G.; Capranica, L.; Guidetti, L.; Pulejo, C. Assessment of static balance in children. *J. Sports Med. Phys. Fit.* **1991**, *31*, 235–242.
31. Riach, C.L.; Starkes, J.L. Velocity of centre of pressure excursions as an indicator of postural control systems in children. *Gait Posture* **1994**, *2*, 167–172. [CrossRef]
32. Rival, C.; Ceyte, H.; Olivier, I. Developmental changes of static standing balance in children. *Neurosci. Lett.* **2005**, *376*, 133–136. [CrossRef]
33. Verbecque, E.; Vereeck, L.; Hallemans, A. Postural sway in children: A literature review. *Gait Posture* **2016**, *49*, 402–410. [CrossRef]
34. Assaiante, C. Action and representation of action during childhood and adolescence: A functional approach. *Clin. Neurophysiol.* **2012**, *42*, 43–51. [CrossRef]
35. Sabharwal, S.; Zhao, C. The Hip-Knee-Ankle Angle in Children: Reference Values Based on a Full-Length Standing Radiograph. *J. Bone Joint Surg. Am.* **2009**, *91*, 2461–2468. [CrossRef]
36. Kernell, D. The final common pathway in postural control developmental perspective. *Neurosci. Biobehav. Rev.* **1998**, *22*, 479–484. [CrossRef]
37. Samsom, J.; DeGroot, L.; Bezemer, P.; Lafeber, H.; Fetter, W. Muscle power development during the first year of life predicts neuromotor behaviour at 7 years in pretermborn. *Early Hum. Dev.* **2002**, *68*, 103–118. [CrossRef]
38. Parfitt, A.M.; Travers, R.; Rauch, F.; Glorieux, F.H. Structural and cellular changes during bone growth in healthy children. *Bone* **2000**, *27*, 487–494. [CrossRef]
39. Kannus, P.; Haapasalo, H.; Sankelo, M.; Sievanen, H.; Pasanen, M.; Heinonen, A.; Oja, P.; Vuori, I. Effect of Starting Age of Physical Activity on Bone Mass in the Dominant Arm of Tennis and Squash Players. *Ann. Int. Med.* **1995**, *123*, 27–31. [CrossRef] [PubMed]
40. Fagard, J. Droitiers/Gauchers: Des asymmetries dans tous les sens. De Boeck superieur Solal: Marseille, France, 2004.
41. Brown, E.R.; Taylor, P.J. Handness, footness and eyeness. *Percept. Mot. Skills* **1988**, *66*, 183–186. [CrossRef]
42. Gilberta, G.; Nassogneb, M. Mise au point face à l'acquisition d'une latéralité précoce de l'enfant. Focus on early acquisition of lateralization in children. *Kinesither. Rev.* **2020**, *20*, 25–28.
43. Massion, J.; Popov, K.; Fabre, J.C.; Rage, P.; Gurfinkel, V. Is the erect posture in microgravity based on the control of trunk orientation or center of mass position? *Exp. Brain Res.* **1997**, *114*, 384–389. [CrossRef]
44. Assaiante, C.; Barlaam, F.; Cignetti, F.; Vaugoyeau, M. Body schema building during childhood and adolescence: A neurosensory approach. *Neurophysiol. Clin.* **2014**, *44*, 3–12. [CrossRef]
45. Naito, E.; Morita, T.; Amemiyaa, K. Body representations in the human brain revealed by kinesthetic illusions and their essential contributions to motor control and corporeal awareness. *Neurosci. Res.* **2016**, *104*, 16–30. [CrossRef] [PubMed]
46. Brown, G.T. The intrinsic factors in the act of progression in the mammal. *Proc. R. Soc. Lond.* **1911**, *84*, 309–319.
47. Brown, G.T. On the nature of the fundamental activity of the nervous centres; together with an analysis of the conditioning of rhythmic activity in progression, and a theory of the evolution of function in the nervous system. *J. Physiol.* **1914**, *48*, 18–46. [CrossRef]
48. Grillner, S.; Zangger, P. Locomotor movements generated by the deafferented spinal cord. *Acta Physiol. Scand* **1974**, *91*, 38A–39A.
49. Grillner, S.; Zangger, P. On the central generation of locomotion in the low spinal cat. *Exp. Brain Res.* **1979**, *34*, 241–261. [CrossRef] [PubMed]
50. Jankowska, E.; Jukes, M.; Lund, S.; Lundberg, A. The effect of DOPA on the spinal cord. 5. Reciprocal organization of pathways transmitting excitatory action to alpha motoneurones of flexors and extensors. *Acta Physiol. Scand.* **1967**, *70*, 369–388. [CrossRef] [PubMed]
51. Jankowska, E.; Jukes, M.; Lund, S.; Lundberg, A. The effect of DOPA on the spinal cord. 6. Half-centre organization of interneurones transmitting effects from the flexor reflex afferents. *Acta Physic. Scand.* **1967**, *70*, 389–402. [CrossRef] [PubMed]
52. Guertin, P. Central pattern generator for locomotion: Anatomical, physiological, and pathophysiological considerations. *Front. Neurol.* **2013**, *3*, 183. [CrossRef]
53. Rowland, R. *Principle of Human Locomotion*; Cambridge Scholars Publishing: Newcastle upon Tyne, UK, 2020.
54. Steuer, I.; Guertin, P. Central pattern generators in the brainstem and spinal cord: An overview of basic principles, similarities, and differences. *Rev. Neurosci.* **2019**, *30*, 107–164. [CrossRef]
55. Tan, U. Two families with quadrupedalism, mental retardation, no speech, and infantile hypotonia (Uner Tan Syndrome Type-II); a novel theory for the evolutionary emergence of human bipedalism. *Front Neurosci.* **2014**, *8*, 84. [CrossRef]

56. Guertin, P.; Dubuc, R. Effects of stimulating the reticular formation during fictive locomotion in lampreys. *Brain Res.* **1996**, *753*, 328–334. [CrossRef]
57. Wannier, T.; Deliagina, G.; Orlovsky, G.; Grillner, S. Differential effects of the reticular system on locomotion in lamprey. *J. Neurophysiol.* **1998**, *80*, 103–112. [CrossRef] [PubMed]
58. Shik, M.L.; Severin, F.V.; Orlovsky, G.N. Control of walking and running by means of electrical stimulation of the midbrain. *Biofizika* **1966**, *11*, 659–666.
59. Shik, M.L.; Severin, F.V.; Orlovsky, G.N. Control of walking and running by means of electrical stimulation of the mesencephalon. *Electroencephalogr. Clin. Neurophysiol.* **1969**, *26*, 549. [PubMed]
60. Forssberg, H.; Grillner, S.; Halbertsma, J.; Rossignol, S. The locomotion of the low spinal cat. II. Interlimb coordination. *Acta Physiol. Scan.* **1980**, *108*, 283–295. [CrossRef]
61. Frigon, A.; Desrochers, É.; Thibaudier, Y.; Hurteau, M.; Dambreville, C. Left-right coordination from simple to extreme conditions during split-belt locomotion in the chronic spinal adult cat. *J. Physiol.* **2017**, *585*, 341–361. [CrossRef]
62. Forssberg, H.; Grillner, S.; Rossignol, S. Phase-dependent reflex reversal during walking in chronic spinal cats. *Brain Res.* **1975**, *85*, 103–107. [CrossRef]
63. Guertin, P.; Angel, M.; Perreault, M.-C.; McCrea, D. Ankle extensor group I afferents excite extensors throughout the hindlimb during fictive locomotion in the cat. *J. Physiol.* **1995**, *487*, 197–209. [CrossRef]
64. Perreault, M.-C.; Angel, M.; Guertin, P.; Mccrea, D. Effects of stimulation of hindlimb flexor group II afferents during fictive locomotion in the cat. *J. Physiol.* **1995**, *487*, 211–220. [CrossRef]
65. Griffin, M.; Olney, S.; McBride, I. Role of symmetry in gait performance of stroke subjects with hemiplegia. *Gait Posture* **1995**, *3*, 132–142. [CrossRef]
66. Kagawa, T.; Nomura, T.; Kondo, S. Interlimb Parallel-link Powered Orthosis (IPPO): Compact Wearable, Robot with Lateral Weight Bearing Mechanisms for Gait Assistance. *IEEE Trans. Med. Robot. Bionics* **2020**, *2*, 300–308. [CrossRef]
67. Meng, S.; Jin, S.; Li, J.; Hashimoto, K.; Guo, S.; Dai, S. The Analysis of HumanWalking Stability Using ZMP in Sagittal Plane. In Proceedings of the IEEE Intertional Conference on Cybernetics and Intelligent Systems, Ningbo, China, 19–21 November 2017; pp. 296–501.
68. Ng, K.; Mehdizadeh, S.; Laboni, A.; Mansfield, A.; Flint, A.; Taati, B. Measuring gait variables using computer vision to assess mobility and fall risk in older adults with dementia. *IEEE J. Transl. Eng. Health Med.* **2020**, *8*, 2100609. [CrossRef]
69. Rodano, R.; Santambrogio, G. Quantitative comparison of locomotor performance in different race walkers. In Proceedings of the 5th International Symposium on Biomechanics in Sports. Conference Proceeding Archive, Athens, Greece, 1987; pp. 122–134.
70. Chodera, J.D.; Levell, R.W. Footprint Patterns during Walking. In *Perspectives in Biomedical Engineering: Proceedings of a Symposium Organised in Association with the Biological Engineering Society and Held in the University of Strathclyde, Glasgow, June 1972*; Kenedi, R.M., Ed.; Palgrave Macmillan: London, UK, 1973; pp. 81–90.
71. Gundersen, L.; Valle, D.; Barr, A.; Danoff, J.; Stanhope, S.; Snyder-Mackler, L. Bilateral analysis of the knee and ankle during gait: An examination of the relationship between lateral dominance and symmetry. *Phys. Ther.* **1989**, *69*, 640–650. [CrossRef] [PubMed]
72. Reisman, D.; Wityk, R.; Silver, K.; Bastian, A. Locomotor adaptation on a split-belt treadmill can improve walking symmetry post-stroke. *Brain* **2007**, *130*, 1861–1872. [CrossRef] [PubMed]
73. Forczek, W.; Staszkiewicz, R. An evaluation of symmetry in the lower limb joints during the able-bodied gait of women and men. *J. Hum. Kinetics* **2012**, *35*, 47–57. [CrossRef]
74. Hannah, R.; Morrison, J.; Chapman, A. Kinematic symmetry of the lower limbs. *Arch. Phys. Med. Rehabil.* **1984**, *65*, 155–158. [PubMed]
75. Karamanidis, K.; Arampatzis, A.; Bruggemann, G. Symmetry and reproducibility of kinematic parameters during various running techniques. *Med. Sci. Sports Exerc.* **2003**, *35*, 1009–1016. [CrossRef] [PubMed]
76. Law, H. Microcomputer-based low-cost method for measurement of spatial and temporal parameters of gait. *J. Biomed. Eng.* **1987**, *9*, 115–120. [CrossRef]
77. Hamill, J.; Bates, B.; Knutzen, K. Ground reaction force symmetry during walking and running. *Res. Q. Exerc. Sport* **1984**, *55*, 289–293. [CrossRef]
78. Herzog, W.; Nigg, B.; Read, L.; Olsson, E. Asymmetries in ground reaction force patterns in normal human gait. *Med. Sci. Sports Exerc.* **1989**, *21*, 110–114. [CrossRef]
79. Hsiao-Wecksler, E.; Polk, J.; Rosengren, K.; Sosnoff, J.; Hong, S. A review of new analytic techniques for quantifying symmetry in locomotion. *Symmetry* **2010**, *2*, 1135–1155. [CrossRef]
80. Arsenault, A.; Winter, D.; Marteniuk, R. Is there a 'normal' profile of EMG activity in gait? *Med. Biol. Eng. Comput.* **1986**, *24*, 337–343. [CrossRef]
81. Õunpuu, S.; Winter, D. Bilateral electromyographical analysis of the lower limbs during walking in normal adults. *Electroencephalogr. Clin. Neurophysiol.* **1989**, *72*, 429–438. [CrossRef]
82. Balakrishan, S.; Thornton-Trump, A. Integral parameters in human locomotion. In Proceedings of the Process 2nd Biannual Conference of the Canadian Society for Biomechanics, Kingston, ON, Canada, 31 August–3 September 1982; pp. 12–13.
83. Damholt, V.; Termansen, N. Asymmetry of plantar flexion strength in the foot. *Acta Orthop.* **1978**, *49*, 215–219. [CrossRef]
84. Vaughan, C. Are joint torques the Holy Grail of human gait analysis? *Hum. Mov. Sci.* **1996**, *15*, 423–443. [CrossRef]

85. Crowe, A.; Schiereck, P.; De Boer, R.; Keessen, W. Characterization of human gait by means of body center of mass oscillations derived from ground reaction forces. *IEEE Trans. Biomed. Eng.* **1995**, *42*, 293–303. [CrossRef]
86. Giakas, G.; Baltzopoulos, V. Time and frequency domain analysis of ground reaction forces during walking: An investigation of variability and symmetry. *Gait Posture* **1997**, *5*, 189–197. [CrossRef]
87. Rice, J.; Seeley, M. An investigation of lower-extremity functional asymmetry for non-preferred able-bodied walking speeds. *Int. J. Exerc. Sci.* **2010**, *3*, 182–188. [PubMed]
88. Seeley, M.; Umberger, B.; Shapiro, R. A test of the functional asymmetry hypothesis in walking. *Gait Posture* **2008**, *28*, 24–28. [CrossRef] [PubMed]
89. Carpes, F.; Mota, C.; Faria, I. On the bilateral asymmetry during running and cycling: A review considering leg preference. *Phys. Ther. Sport* **2010**, *11*, 136–142. [CrossRef] [PubMed]
90. Hall, C.; Figueroa, A.; Fernhall, B.; Kanaley, J.A. Energy expenditure of walking and running: Comparison with prediction equations. *Med. Sci. Sports Exerc.* **2004**, *36*, 2128–2134. [CrossRef]
91. Sardian, P.; Bessonnet, G. Forces acting on a biped robot. Center of pressure-zero moment point. *Trans. Syst. Man Cybern. Part A Syst. Hum.* **2004**, *34*, 630–637. [CrossRef]
92. Vukobratovic, M.; Borovac, B. Zero-moment point–thirty five years of its life. *Int. J. Hum. Robot.* **2004**, *1*, 157–173. [CrossRef]
93. Vukobratovic, M.; Frank, A.; Juricic, D. On the stability of biped locomotion. *IEEE Trans. Biomed. Eng.* **1970**, *BME-17*, 25–36. [CrossRef]
94. Wada, Y.; Kanekoa, Y.; Nakano, E.; Osu, R.; Kawato, M. Quantitative examinations for multi joint arm trajectory planning using a robust calculation algorithm of the minimum commanded torque change trajectory. *Neural Netw.* **2001**, *14*, 381–393. [CrossRef]
95. Li, Q.; Takanishi, A.; Kato, I. Learning control for a biped walking robot with a trunk. In Proceedings of the Process of 1993 IEEE-RSJ International Conference on Intelligent Robots and Systems IROS 93, Yokohama, Japan, 26–30 July 1993.
96. Fujimoto, Y.; Kawamura, A. Three dimensional digital simulation and autonomous walking control for eight-axis biped robot. In Proceedings of the IEEE Intenational Conference on Robotics and Automation, Nagoya, Japan, 21–27 May 1995; pp. 2877–2884.
97. Hirai, K.; Hirose, M.; Haikawa, Y.; Takenaka, T. The development of Honda humanoid robot. In Proceedings of the IEEE International Conference on Robotics and Automation (ICRA), Leuven, Belgium, 16–20 May 1998; pp. 1321–1326.
98. Takanishi, A.; Ishida, M.; Yamazaki, Y.; Kato, I. The realization of dynamic walking by the biped walking robot wl-10rd. In Proceedings of the IEEE International Conference on Robotics and Automation, St. Louis, MO, USA, 25 March 1985; pp. 459–466.
99. Nishiwaki, K.; Kagami, S.; Kuniyoshi, Y.; Inaba, M.; Inoue, H. Online generation of humanoid walking motion based on a fast generation method of motion pattern that follows desired ZMP. In Proceedings of the IEEE/RSJ International Conference on Intelligent Robots and Systems, Lausanne, Switzerland, 30 September–4 October 2002; pp. 2684–2689.
100. Huang, Q.; Peng, Z.; Zhang, W.; Zhang, L.; Li, K. Design of humanoid complicated dynamic motion based on human motion capture. In Proceedings of the IEEE/RSJ International Conference on Intelligent Robots and Systems, Edmonton, AB, Canada, 2–6 August 2005.
101. Gorce, P. Dynamic postural control method for biped in unknown environment. *IEEE Trans. Syst. Man Cybern.* **1999**, *29*, 616–626. [CrossRef]
102. Gorce, P.; Guihard, M. On dynamics control of pneumatic bipeds. *J. Robotic Syst.* **1998**, *15*, 421–433. [CrossRef]
103. Gorce, P.; Hafi, E.l.; Coronado, J.L. Dynamic control of walking cycle with initiation process for humanoid robot. *J. Intell. Robotics Syst.* **2001**, *31*, 321–337. [CrossRef]
104. Gorce, P.; Vanel, O.; Ribreau, C. Equilibrium study of human robot. In Proceedings of the IEEE International Conference on Systems Man and Cybernetics, Vancouver, BC, Canada, 22–25 October 1995; pp. 1309–1314.
105. Hirukawa, H.; Hattori, S.; Kajita, S.; Harada, K.; Kaneko, K.; Kanehiro, F.; Morisawa, M.; Nakaoka, S. A pattern generator of humanoid robots walking on a rough terrain. In Proceedings of the 2007 IEEE International Conference on Robotics and Automation, Roma, Italy, 10–14 April 2007; pp. 2181–2187.
106. Kim, S.; Murakami, T. An Approach to Human Walking Analysis Based on Balance, Symmetry and Stability Using COG, ZMP and CP. *Appl. Sci.* **2020**, *10*, 7307. [CrossRef]
107. Goswami, A. Postural Stability of Biped Robots and the Foot-Rotation Indicator (FRI) Point. *Int. J. Robot. Res.* **1999**, *18*, 523–533. [CrossRef]
108. Ortiz, A.; Ibarra, J. Walk stability control for position-controlled servo actuated humanoid robot. In Proceedings of the IEEE International Conference on Electrical Engineering, Computing Science and Automatic Control, Mexico City, Mexico, 11–13 September 2019; pp. 1–6.
109. Raoufi, M.; Edrisi, M. Designing a Fractional Order Back-Stepping Controller Based on GPI Observer for a 3D Biped Robot. In Proceedings of the IEEE International Conference on Robotics and Mechatronics, Tehran, Iran, 20–21 November 2019; pp. 211–216.
110. Soliman, A.; Sendur, P.; Ugurlu, B. 3-D DynamicWalking Trajectory Generation for a Bipedal Exoskeleton with Underactuated Legs: A Proof of Concept. In Proceedings of the IEEE International Conference on Rehabilitation Robotics, Toronto, CN, Canada, 24–28 June 2019; pp. 599–604.
111. Tesio, L.; Rota, V. The motion of body center of mass during walking: A review oriented to clinical applications. *Front. Neurol.* **2019**, *10*, 1–22. [CrossRef] [PubMed]

symmetry

MDPI

Article

Cerebral Arterial Asymmetries in the Neonate: Insight into the Pathogenesis of Stroke

Anica Jansen van Vuuren [1], Michael Saling [1,2,*], Sheryle Rogerson [3], Peter Anderson [4,5], Jeanie Cheong [3,6] and Mark Solms [7]

1 Melbourne School of Psychological Sciences, University of Melbourne, Melbourne, VIC 3010, Australia; anicajvv@gmail.com
2 Florey Institute of Neurosciences and Mental Health, Royal Parade, Melbourne, VIC 3052, Australia
3 Royal Women's Hospital, Flemington Rd., Melbourne, VIC 3052, Australia; Sheryle.Rogerson@thewomens.org.au (S.R.); Jeanie.Cheong@thewomens.org.au (J.C.)
4 Murdoch Children's Research Institute, Royal Children's Hospital, Flemington Rd., Melbourne, VIC 3052, Australia; peter.anderson@mcri.edu.au
5 Turner Institute for Brain and Mental Health, Monash University, Melbourne, VIC 3800, Australia
6 Clinical Sciences, Murdoch Children's Research Institute, Flemington Road, Melbourne, VIC 3052, Australia
7 Department of Psychology, University of Cape Town, Private Bag, Cape Town 7701, South Africa; mark.solms@uct.ac.za
* Correspondence: mmsaling@unimelb.edu.au

Abstract: Neonatal and adult strokes are more common in the left than in the right cerebral hemisphere in the middle cerebral arterial territory, and adult extracranial and intracranial vessels are systematically left-dominant. The aim of the research reported here was to determine whether the asymmetric vascular ground plan found in adults was present in healthy term neonates (n = 97). A new transcranial Doppler ultrasonography dual-view scanning protocol, with concurrent B-flow and pulsed wave imaging, acquired multivariate data on the neonatal middle cerebral arterial structure and function. This study documents for the first-time systematic asymmetries in the middle cerebral artery origin and distal trunk of healthy term neonates and identifies commensurately asymmetric hemodynamic vulnerabilities. A systematic leftward arterial dominance was found in the arterial caliber and cortically directed blood flow. The endothelial wall shear stress was also asymmetric across the midline and varied according to vessels' geometry. We conclude that the arterial structure and blood supply in the brain are laterally asymmetric in newborns. Unfavorable shearing forces, which are a by-product of the arterial asymmetries described here, might contribute to a greater risk of cerebrovascular pathology in the left hemisphere.

Keywords: middle cerebral artery; diameter; blood flow; asymmetry; stroke; shear stress; neonate

Citation: van Vuuren, A.J.; Saling, M.; Rogerson, S.; Anderson, P.; Cheong, J.; Solms, M. Cerebral Arterial Asymmetries in the Neonate: Insight into the Pathogenesis of Stroke. *Symmetry* **2022**, *14*, 456. https://doi.org/10.3390/sym14030456

Academic Editor: Thierry Paillard

Received: 22 December 2021
Accepted: 24 January 2022
Published: 24 February 2022

Publisher's Note: MDPI stays neutral with regard to jurisdictional claims in published maps and institutional affiliations.

1. Introduction

Middle cerebral artery strokes occur more commonly in the left cerebral hemisphere [1,2]. In the mature brain, this leftward predilection has been attributed by some to selective recognition of the clinically obvious sequalae of left hemispheric events [3]. Neurovascular vulnerabilities that might explain a left hemispheric predilection for stroke have also been identified [4].

Adult studies report left-biased asymmetries in the structure and hemodynamics of extracranial and intracranial arteries, namely, the vertebral arteries [5,6], common and internal carotid arteries [7], and middle and anterior cerebral arteries [8]. These reports of larger arterial calibers, higher flow velocities, and higher blood flow volumes on the left are in keeping with the notion of a more resource intensive left hemisphere [9] and create left–right differences in the circulations of each arterial tree.

Hemodynamic processes, such as changes in blood pressure parameters, the speed of the pressure wave propagation, and resulting shearing forces on the arterial endothelium, play important roles in the development of vascular disease [10]. The distribution of atherosclerosis in the vascular system is not uniform and plaque severity and composition also varies according to location [11]. Reports of higher left-than-right intima-media wall thickness [12] and plaque incidence, thickness, and instability [3] in the carotid arteries suggest a lateralized vulnerability to cerebrovascular disease in adults.

A left hemisphere predilection for cerebrovascular pathology, such as periventricular hemorrhage [13], neonatal stroke [14], and cerebral palsy [15], has also been reported in neonates. Approximately 70% to 80% of neonatal ischemic strokes occur in the middle cerebral arterial field, and they are left-sided in 53% to 75% of cases [16]. This begs a key question: is the ground plan of adult arterial asymmetries and corresponding vulnerability to pathology discernible in neonates? There is only one study, to our knowledge, that aimed to investigate the significance of left–right differences in the blood flow velocity to neonatal stroke, but only 20 normal control cases were reported, without data on arterial diameter, flow volume, or shear stress [17].

Ultrasonography of neonatal cerebral arteries is common in routine clinical practice and largely proceeds on the assumption of trans-midline symmetry. The resolution of existing methodologies has not been extended to detect the existence of structurofunctional asymmetries]. In previously used Doppler technologies, "bleeding", blooming artefact, and the influence of gain settings are recognized sources of error, particularly in relation to diametric measurement. This is problematic, since conclusions about regional cerebral blood flow cannot be drawn from velocity measurements [18], primarily because the volume flow (Q) in a vessel is related to the velocity (V) as well as the vessel's radius (R) according to the equation $Q = V\pi R^2$. Similarly, the calculation of the wall shear stress requires a diametric measurement according to the equation $\tau = 4\mu(V/\pi D^3)$. B-flow imaging is a recently introduced non-Doppler technology which effectively bypasses these difficulties [19]. A dual-view imaging protocol using concurrent pulsed wave and B-flow Doppler transcranial ultrasonography addresses these shortcomings and paves the way for investigating the aims of the research reported here, namely, to investigate neonatal arterial asymmetry and corresponding cerebrovascular vulnerabilities.

We focused on the trunk of the middle cerebral artery as a major and accessible conduit to the lateral neocortical territory. We hypothesized that the diameter, hemodynamics, and shear stress are all inherently asymmetric in the direction of larger arterial calibers, higher blood flow volumes, and unfavorable shear stress on the left in the majority of healthy term neonates.

2. Materials and Methods

2.1. Search Strategy and Selection Criteria

Transcranial Doppler ultrasonography was performed on 106 healthy term neonates. Neonates with a gestational age greater than 37 weeks were recruited consecutively between March 2017 and November 2017 from the postnatal wards of the Royal Women's Hospital and Frances Perry House in Melbourne, Australia. A non-randomized participant sampling approach accompanied by comprehensive exclusion criteria (see below) was adopted. Six participants were excluded from the final analysis because aberrant middle cerebral arterial branching patterns precluded left–right comparisons of arterial geometry and hemodynamics. An additional three participants were excluded for poor image quality because of excessive neonatal movement, excessive hair, and or small cranial windows.

2.2. Neonatal Exclusion Criteria

Infants with significant perinatal complications were excluded (for example, postnatal resuscitation and/or admission to the neonatal intensive and special care nursery). Neonates with an intracranial pathology, substance exposure, or metabolic, genetic, and/or cardiovascular disorders were excluded. All infants enrolled in the study were healthy and had no dysmorphic features during the neonatal predischarge check.

2.3. Maternal Exclusion Criteria

Exclusion criteria included diagnoses of autoimmune disorders, pre-gestational diabetes mellitus, gestational diabetes, cardiac disease, drug and substance use, instances of suspected or detected fetal abnormality prior to delivery, chronic or persistent hypertension (>140/90), infections (including active genital herpes, syphilis, and HIV+), pre-eclampsia, and neurological and mental health conditions. Non-English-speaking parents were excluded from the study to ensure effective communication and understanding between the parent and investigators.

All scanning took place at the Royal Women's Hospital, Melbourne, Australia. Ethical approval was granted by the Royal Women's Hospital Human Research Ethics Committee and written informed consent was obtained from one or both parents.

2.4. Procedure

Transcranial ultrasonography and Doppler assessment took place at a postnatal age of 1 to 7 days. Scans did not reflect acute hemodynamic changes known to occur in the first 12 h of life [20]. Standard medical procedure was followed prior to the analysis. All infants underwent 10 min of supine rest on a clean cot in a standardized sound proofed ultrasound room with no auditory or visual distractions. The room had constant illumination and comfortable room temperature. Neonates were swaddled and fed prior to the scanning session. Parents were positioned at the head of the cot, behind the investigator so as not to distract the infant. If the neonate began to cry, the neonate was soothed before resuming the procedure.

Transcranial Doppler cerebrovascular imaging was performed using the portable General Electric EPIQ 9 ultrasound unit (GE Healthcare, Wauwatosa, WI, USA). A C3-10-D convex probe (2–11 MHz) with an insonation angle close to 0° was used. Further settings included a small sample volume of 2 mm with a velocity wall filter of 80–100 Hz to eliminate noise caused by vessel wall movement.

Using a trans-temporal approach, the middle cerebral artery trunk was located by placing the transducer on the left temporal bone, below the zygomatic arch. Screening for previously undetected pathology and identification of the middle cerebral artery was performed with two-dimensional B mode gray-scale imaging and color flow imaging through the temporal window. B-flow imaging was activated, and the probe was moved so as to optimize the visualization of the origin of the middle cerebral artery trunk (approximately 2 mm from internal carotid artery terminus). At this distance, the vessel has a uniform diameter and required minimal angle correction. In any necessary instance, an angle of correction was performed if the angle of incidence was greater than 15° to ensure the transducer remained parallel to the vector of blood flow and accurate measures were obtained. Dual-view imaging was then initiated to replicate the image into two identical left and right images. The left image was selected, pulsed wave Doppler was activated, and several hemodynamic measurements were recorded at the arterial site. B-flow and pulsed wave frequencies used were 6.0 MHz and 4.2 MHz respectively. Three distinct pulsed wave spectral tracings containing three consecutive cardiac cycles were recorded. Peak systolic velocity (PSV), end-diastolic velocity (EDV), time averaged maximum velocity (TA_{MAX}), time averaged mean velocity (TA_{MEAN}), and heart rate measures were obtained. An on-site arterial diameter was taken from the corresponding right B-Flow image in the exact location hemodynamic measures were sourced.

The distal portion of the middle cerebral artery trunk (distal to the origins of the lenticulostriate arteries) approximately 2 mm from the middle cerebral artery bifurcation/trifurcation was located, and hemodynamic and diameter measures were repeated. The procedure was then repeated on the contralateral Mo and M_{DT} sites. The sequence of data collection from the left and right middle cerebral arteries was randomized.

As a proof of concept for the new scanning protocol, we also imaged the very fine lenticulostriate branches of the middle cerebral trunk to a high degree of resolution. Lenticulostriate artery sampling in the study was sparse, largely because these vessels are difficult to image and of a small caliber. The fact that lenticulostriate arteries were imaged to the point of supporting reliable measurement attests to the resolution of the innovations that were introduced to accomplish this.

The lenticulostriate arteries of the left and right cerebral hemisphere were approached by placing the same C3-10 transducer in the mid-sagittal plane of the anterior fontanelle. The transducer was fanned into the left cerebral hemisphere. Screening and identification of the lenticulostriate arteries was performed with two-dimensional B mode gray-scale imaging and color flow imaging. B-flow imaging was activated and two lenticulostriate arteries in each cerebral hemisphere were chosen for further scanning based on the clarity of the image and orientation of the vessel (that is, the two arteries on each side that were most oriented in the vertical plane). The probe was moved so as to optimize the visualization of one of the selected vessels. B-flow and pulsed wave Doppler was utilized in dual-view imaging to record structural and hemodynamic measures of the lenticulostriate artery. The procedure was then repeated for the second unilateral and two contralateral lenticulostriate arteries in a randomized order.

All images were stored on optical disc for off-line analysis using SYNAPSE (PACS) 64-bit imaging software. All hemodynamic measures were averaged across three homogenous consecutive cardiac cycles for each arterial site. Further investigation of arterial diameter was performed offline with RadiAnt DICOM viewer (64-bit) imaging software (version 4.2.1). The mean lumen diameter of each arterial site was determined by averaging three independent measurements taken at the same location as on-line analyses. Parameters were also averaged across the ipsilateral origin and distal trunk of middle cerebral artery (MCA_{MEAN}). Assessment of inter-rater reliability was performed by SR, an experienced sonographer, on 10% of participants randomly selected from the sample throughout the data collection period. Cronbach's alpha showed a high internal consistency of 0.963.

At each site, the following hemodynamic indices were calculated with the following formulae.

Mean velocity (V_{MEAN}):

$$V_{MEAN} = \frac{PSV + EDV}{2}$$

Resistive index (RI) computed according to the method of Pourcelot (1982):

$$RI = \frac{PSV - EDV}{PSV}$$

Pulsatility index (PI) computed according to the method of Gosling and King (1988):

$$PI = \frac{PSV - EDV}{V_{MEAN}}$$

Shear stress (τ; dyne/cm^2):

$$\tau = 4 \cdot \mu \cdot \frac{V}{\pi \cdot D^3}$$

where V equals the flow velocity, μ equals the viscosity of flow, and D equals the arterial diameter. No data was available concerning blood viscosity in the neonates, so an average neonatal hematocrit-adjusted (0–45) blood viscosity, adjusted at high shear rates of 4.22 mPa.s, was assumed [21], as there is no reason to suspect intraindividual viscosity differences or systematic differences between left- and right-dominant neonates.

Volume flow (Q):

$$Q = PSV \times \left(D^2\left(\frac{\pi}{4}\right)\right)$$

Peak systolic velocity was used as a variable in the calculation of volume flow because it is sensitive to left–right differences in the neonate [22], is mediated by arterial structure [23], and reflects cerebral blood flow [24], the definition of which is the primary aim of this work. Average measures (such as TA_{MEAN}) inevitably conflate peak systolic velocity with end-diastolic velocity. While this might be useful in particular clinical applications, end-diastolic velocities show less left–right differentiation [25].

Arterial diameter was used as a grouping variable for the sample. Interhemispheric diameter dominance was expressed in the form of a left–right laterality index and calculated with the formula:

$$Laterality\ index = \frac{L - R}{L + R}$$

where R equals the right arterial measure and L the left arterial measure. A positive value indicated left arterial dominance, whereas a negative value indicated right arterial dominance. A score of 0 represents the absence of a structural dominance. A LI was calculated for each arterial site as well as the cerebral artery average between the middle cerebral origin and distal trunk (MCA_{MEAN}).

2.5. Statistical Analyses

Data were analyzed using IBM SPSS Statistics (version 23) software. Each hemodynamic measure of the middle cerebral arteries was analyzed using a mixed-design ANOVA. Neonates with no structural arterial dominance were removed from the analysis. For each analysis, the within-subjects factor was the respective arterial parameter (of the left and right paired arteries) and the between-subjects factor was the structural dominance (left-dominant or right-dominant). One-tailed paired t-tests compared lateral differences in geometric groups in instances of significant interactions. One-tailed independent t-tests also compared sex differences in participant demographics and hemodynamics parameters at each site of measurement. Tests of normality and homoscedasticity (namely Levene's test of equality of variance and Shapiro–Wilk tests) were run on each dataset. If the assumption of normality was not upheld, a non-parametric Mann–Whitney U test was run instead.

For brevity, results for TA_{MAX} and TA_{MEAN} are omitted as they are collinear with PSV ($r > 0.90$) and V_{MEAN} ($r > 0.90$), respectively, as described below, but might not be as precise as PSV in defining lateral difference (see above).

Internal consistency was calculated using Cronbach's alpha in 10% of cases. Cohen's rule of thumb for effect size interpretations was used for between-group comparisons: $d = 0.10$ (small effect), $d = 0.30$ (medium effect), and $d = 0.50$ (large effect). The significance of the analyses was determined with a 95% confidence level at $p < 0.05$.

3. Results

The geometric and hemodynamic properties of the middle cerebral artery origin and termination of its trunk were recorded in 97 healthy full-term neonates. The final sample included 59 males and 38 females born via normal vaginal delivery or caesarean section (Table 1). The gestational age at birth of the sample ranged from 36 to 41 weeks, and birth weights ranged from 2200 g to 4930 g. The postnatal age at the time of scanning was 12 to 174 h ($M = 47.71$ h; $SD = 28.58$; $Median = 41$ h; $Range = 162$ h). Mean Apgar scores were 8.30 at 1 min ($SD = 1.38$) and 8.92 at 5 min ($SD = 0.32$). The sample spent an average of 68 h in hospital. There were no significant sex differences in the birth weight, postnatal scanning age, or gestational age.

As an example of dual views of B-mode and pulsed wave Doppler ultrasound imaging, Figure 1 illustrates the diameter and hemodynamic variability of the left and right middle cerebral arteries. As a proof of concept, we also imaged the very fine lenticulostriate branches of the middle cerebral trunk to a high degree of resolution. Example images are included in Figure 1. Demographic information on the neonates according to their averaged middle cerebral geometric asymmetry (left-dominant or right-dominant) is presented in Table 1.

Table 1. Neonatal characteristics as a function of geometric arterial asymmetry.

	Left-Dominant	Right-Dominant	No Dominance	Total
	M (SD)	*M (SD)*	*M (SD)*	*M (SD)*
N [a]	57	34	6	97
Sex (%)				
Male	61.4	64.7	66.7	62.9
Female	39.6	35.3	33.3	37.1
Gestational age at birth (wk)	39.07 (1.45)	38.76 (1.46)	39.00 (1.67)	38.99 (1.45)
Age at scan (hrs)	48.44 (32.17)	48.88 (23.94)	38.00 (11.22)	47.71 (28.58)
Birth weight (g)	3418.51 (539.94)	3525.15 (621.10)	3428.00 (470.34)	3460.76 (560.37)
AS^{1min}	8.40 (1.31)	8.18 (1.49)	8.5 (1.22)	8.30 (1.38)
AS^{5min}	8.91 (0.39)	8.97 (0.17)	9 (0.00)	8.92 (0.32)
Heart rate (bt/min)	113.62 (14.64)	111.82 (13.79)	114.58 (17.07)	113.87 (13.89)

Note: AS^{1min} = Apgar score at 1 min; AS^{5min} = Apgar score at 5 min; [a] = number according to geometric dominance averaged across middle cerebral origin and distal trunk.

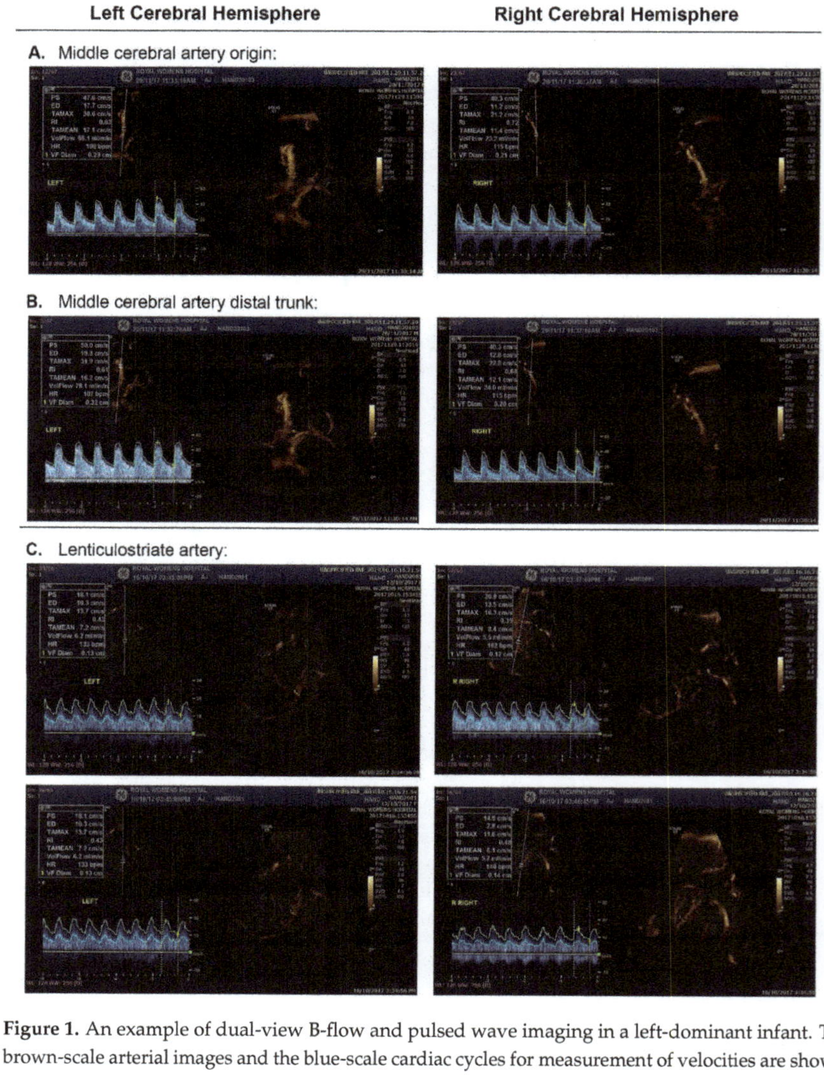

Figure 1. An example of dual-view B-flow and pulsed wave imaging in a left-dominant infant. The brown-scale arterial images and the blue-scale cardiac cycles for measurement of velocities are shown.

Panels (**A**): origin of the left and right middle cerebral arteries. The diameter on the left is 2.9 mm and the diameter on the right is 2.1 mm; the difference is visible on inspection of the brown-scale images. Peak systolic velocity on the left (PS in the quantitative panel) is 47.6 cm/s and 40.3 cm/s on the right. The end-diastolic velocity (ED in the quantitative panel) is 17.7 cm/s on the left and 11.2 cm/s on the right. Panels (**B**): distal segment of the trunk of the left and right middle cerebral arteries. The diameter on the left is 3.2 mm and 2.0 mm on the right, and the difference is again visible on inspection of the brown-scale images. Peak systolic velocity (PS) is 50 cm/s on the left and 40.3 cm/s on the right. End-diastolic velocity (ED) is 19.3 cm/s on the left and 12.8 cm/s on the right. Panels (**C**): the lenticulostriate arteries are shown largely as a proof of the concept that very small caliber arteries in the neonatal brain can be visualized and that structurofunctional measurements can be obtained. The arteries selected for measurement can be identified by the white dotted lines in the brown-scale images.

3.1. Sex Differences

No significant sex differences were found in the arterial diameter; peak systolic, end-diastolic, and mean velocity; and resistance or pulsatility indices. A significant sex difference was found in the blood flow volume in the left middle cerebral origin. Overall, males had higher left-sided blood flow volumes (M = 210.04 mL/min; SD = 75.05) than females (M = 189.87 mm; SD = 80.50) at this arterial site (p = 0.041). Shearing forces at each corresponding arterial site were comparable between males and females apart from the shear stress in the distal trunk of the right middle cerebral artery. Females had a higher right-sided wall shear stress (M = 595.99 dyne/cm^2; SD = 208.90) than males (M = 504.71 dyne/cm^2; SD = 137.61) at the distal trunk (p = 0.034).

3.2. Structural Differences

Left–right asymmetries in the arterial diameter were found at each arterial site ($p < 0.001$; Table 2). Of the 97 participating neonates, a left geometric dominance was exhibited in 52 (54%) at the middle cerebral origin and 60 (62%) at the middle cerebral distal trunk. When averaged across the arteries, with no consideration of individual dominance, a significant leftward structural difference was evident only at the middle cerebral distal trunk ($t(96) = 1.989$, $p = 0.050$, $d = 0.239$). A small proportion of participants showed no left–right differences in the arterial diameter at the origin (8%) and distal trunk (4%). Laterality indices of structure at the middle cerebral artery proximal segment were associated with asymmetries at the distal segment ($r = 0.741$; $p < 0.001$).

Analyses described in this paper have not been undertaken in previous work. Rather, left–right comparisons classically are made on the basis of average values across the entire sample and with measurements taken at a single site, namely the origin of the middle cerebral artery. The findings reported in the "Averaged" column of Table 2 show that this approach hides or reduces the probability of the systematic individual lateral dominance reported here.

3.3. Structurofunctional Differences

Considerable geometric and hemodynamic asymmetries existed in the origin and distal trunk of the middle cerebral artery. In participants with a leftward dominance in the arterial geometry, peak blood flow velocities were higher on the left side at both sites of the middle cerebral artery. A leftward bias in the average flow velocity was found at the origin, and higher blood flow volumes were also found in the larger left origin and distal trunk of this group. No lateral differences in the end-diastolic velocity were found at either site.

Across both middle cerebral arterial sites, no lateral differences in the peak systolic, end-diastolic of average blood flow velocity were found in neonates with larger arteries on the right side (Tables 3 and 4; Figure 2). Converse to neonates with a leftward dominance in the geometry, a right-sided asymmetry in the overall blood flow volume was found at the origin and distal trunk of neonates with larger arteries in the right hemisphere.

Table 2. Intra-individual left–right diametric differences as a function of inter-individual differences in the direction of arterial asymmetry.

		Left-Dominant	Right-Dominant	No Dominance	Averaged [a]
Artery	Side	*M (SD)*	*M (SD)*	*M (SD)*	*M (SD)*
MCA_O (mm)	*n*	52	37	8	97
	L	2.32 (0.35)	1.95 (0.25)	2.01 (0.17)	2.13 (0.35)
	R	1.91 (0.23)	2.39 (0.43)	2.01 (0.17)	2.11 (0.39)
	p	0.000	0.000	-	0.412
MCA_{DT} (mm)	*n*	60	33	4	97
	L	2.14 (0.35)	1.87 (0.27)	2.03 (0.20)	2.04 (0.34)
	R	1.81 (0.22)	2.22 (0.34)	2.03 (0.20)	1.96 (0.33)
	p	0.000	0.000	-	0.050 *
MCA_{MEAN} (mm)	*n*	57	34	6	97
	L	2.14 (0.35)	1.86 (0.27)	2.03 (0.16)	2.09 (0.33)
	R	1.81 (0.22)	2.22 (0.35)	2.03 (0.16)	2.03 (0.34)
	p	0.000	0.000	-	0.158

Note: $p < 0.05$; MCA_O = middle cerebral artery origin; MCA_{DT} = middle cerebral artery distal trunk; MCA_{MEAN} = middle cerebral artery averaged across origin and distal trunk measures; [a] = averaged across the sample with no consideration of individual differences in arterial asymmetry.

Table 3. Comparisons of hemodynamic parameters between left and arterial sites according to geometric dominance.

			Left Hemisphere		Right Hemisphere					
Artery	Dominance	Parameter	*M*	*SD*	*M*	*SD*	t	df	*p*	d
MCA_O	Left	PSV (cm/s)	54.00	10.23	51.24	11.28	2.307	51	0.013 *	0.331
		EDV (cm/s)	18.91	4.42	18.25	5.45	1.140	51	0.130	0.154
		V_{MEAN} (cm/s)	30.58	5.77	29.25	6.92	1.866	51	0.033 *	0.257
		RI	0.65	0.07	0.64	0.07	0.250	51	0.402	0.163
		PI	0.96	0.14	0.96	0.15	0.195	51	0.423	0.000
		Q (mL/min)	232.52	83.44	149.98	41.15	7.509	51	0.000 *	1.048
		WSS_{SYS} (dyne/cm^2)	80.36	18.77	91.32	22.39	−5.118	51	0.000 *	−0.727
		WSS_{DIAS} (dyne/cm^2)	2.80	0.68	3.24	1.00	−2.022	51	0.000 *	−0.683
	Right	PSV (cm/s)	56.04	14.85	57.67	14.62	−0.945	36	0.176	0.155
		EDV (cm/s)	18.63	6.97	19.53	6.28	−1.253	36	0.109	0.230
		V_{MEAN} (cm/s)	31.10	9.17	32.24	8.73	−1.167	36	0.125	0.191
		RI	0.67	0.07	0.66	0.06	0.857	36	0.199	0.187
		PI	1.01	0.15	0.99	0.13	0.937	36	0.178	0.173
		Q (mL/min)	167.15	52.67	266.11	123.28	−5.630	36	0.000 *	0.926
		WSS_{SYS} (dyne/cm^2)	99.51	32.68	84.14	27.73	20.270	36	0.001 *	0.572
		WSS_{DIAS} (dyne/cm^2)	3.32	1.46	2.87	1.19	0.664	36	0.007 *	0.432
MCA_{DT}	Left	PSV (cm/s)	53.26	12.72	51.03	14.23	1.947	59	0.028 *	0.251
		EDV (cm/s)	18.28	5.53	18.14	6.45	0.245	59	0.407	0.032
		V_{MEAN} (cm/s)	29.94	7.52	29.10	8.65	1.173	59	0.123	0.152

Table 3. *Cont.*

Artery	Dominance	Parameter	Left Hemisphere		Right Hemisphere		t	df	p	d
			M	SD	M	SD				
		RI	0.66	0.06	0.65	0.07	1.773	59	0.041 *	0.193
		PI	0.99	0.14	0.96	0.15	1.726	59	0.045 *	0.260
		Q (mL/min)	193.68	71.06	130.20	37.92	8.093	59	0.000 *	1.045
		WSS_{SYS} (dyne/cm^2)	86.69	26.81	97.56	33.08	−4.597	59	0.000 *	−0.628
		WSS_{DIAS} (dyne/cm^2)	2.97	1.06	3.48	1.36	−4.574	59	0.000 *	−0.625
	Right	PSV (cm/s)	53.90	12.42	54.73	11.86	−0.628	32	0.267	0.109
		EDV (cm/s)	17.06	5.52	17.56	4.98	−0.735	32	0.233	0.127
		V_{MEAN} (cm/s)	29.58	7.71	29.95	6.70	−0.422	32	0.338	0.073
		RI	0.68	0.06	0.68	0.06	0.629	32	0.267	0.000
		PI	1.03	0.15	1.03	0.15	0.588	32	0.267	0.000
		Q (mL/min)	148.46	50.63	217.59	88.24	−6.771	32	0.000 *	1.178
		WSS_{SYS} (dyne/cm^2)	100.26	30.39	84.80	20.63	4.705	32	0.000 *	0.940
		WSS_{DIAS} (dyne/cm^2)	3.16	1.22	2.71	0.80	3.081	32	0.002 *	0.602

Note: $p < 0.05$; MCAO = middle cerebral artery origin; MCADT = middle cerebral artery distal trunk; PSV = peak systolic velocity; EDV = end diastolic velocity; VMEAN = mean velocity; RI = resistance index; PI = pulsatility index; Q = blood flow volume; WSSSYS = systolic wall shear stress; WSSDIAS = diastolic wall shear stress; * = significant *p* values.

Table 4. Effects of geometric dominance on hemodynamic parameters of arterial sites.

	Source	Middle Cerebral Origin				Middle Cerebral Distal Trunk			
		df	F	p	ηp^2	df	F	p	ηp^2
PSV	PSV	1	0.266	0.607	0.003	1	0.584	0.447	0.006
	Geometric dominance	1	2.902	0.092	0.032	1	0.663	0.418	0.007
	PSV* Geometric dominance	1	4.621	0.034 *	0.050	1	2.795	0.098	0.030
EDV	EDV	1	0.055	0.815	0.001	1	0.154	0.696	0.002
	Geometric dominance	1	0.184	0.669	0.002	1	0.613	0.436	0.007
	EDV* Geometric dominance	1	2.877	0.093	0.032	1	0.484	0.488	0.005
V_{MEAN}	V_{MEAN}	1	0.028	0.868	0.000	1	0.162	0.689	0.002
	Geometric dominance	1	1.345	0.249	0.015	1	0.023	0.879	0.000
	V_{MEAN}* Geometric dominance	1	4.382	0.039 *	0.048	1	1.079	0.302	0.012
RI	RI	1	0.580	0.448	0.007	1	2.441	0.122	0.026
	Geometric dominance	1	2.390	0.126	0.027	1	4.779	0.031 *	0.050
	RI* Geometric dominance	1	0.175	0.667	0.002	1	0.331	0.566	0.004
PI	PI	1	0.658	0.419	0.008	1	2.242	0.138	0.024
	Geometric dominance	1	2.312	0.132	0.026	1	4.818	0.031 *	0.050
	PI* Geometric dominance	1	0.304	0.583	0.003	1	0.273	0.603	0.003

Table 4. *Cont.*

	Source	Middle Cerebral Origin df	F	p	ηp²	Middle Cerebral Distal Trunk df	F	p	ηp²
Q	Q	1	0.693	0.408	0.008	1	0.188	0.665	0.002
	Geometric dominance	1	3.294	0.073	0.036	1	3.131	0.080	0.033
	Q* Geometric dominance	1	84.636	0.000 *	0.493	1	103.905	0.000 *	0.533
WSS_{SYS}	WSS_{SYS}	1	0.939	0.335	0.011	1	1.303	0.257	0.014
	Geometric dominance	1	1.492	0.225	0.017	1	0.005	0.945	0.000
	WSS_{SYS}* Geometric dominance	1	33.575	0.000 *	0.278	1	43.048	0.000 *	0.321
WSS_{DIAS}	WSS_{DIAS}	1	0.001	0.976	0.000	1	0.046	0.861	0.001
	Geometric dominance	1	0.109	0.742	0.001	1	1.367	0.245	0.015
	WSS_{DIAS}* Geometric dominance	1	22.661	0.000 *	0.207	1	27.122	0.000 *	0.230

Note. $p < 0.05$; PSV = peak systolic velocity; EDV = end diastolic velocity; VMEAN = mean velocity; RI = resistance index; PI = pulsatility index; Q = blood flow volume; WSS_{SYS} = systolic wall shear stress; WSSDIAS = diastolic wall shear stress; * significant p values

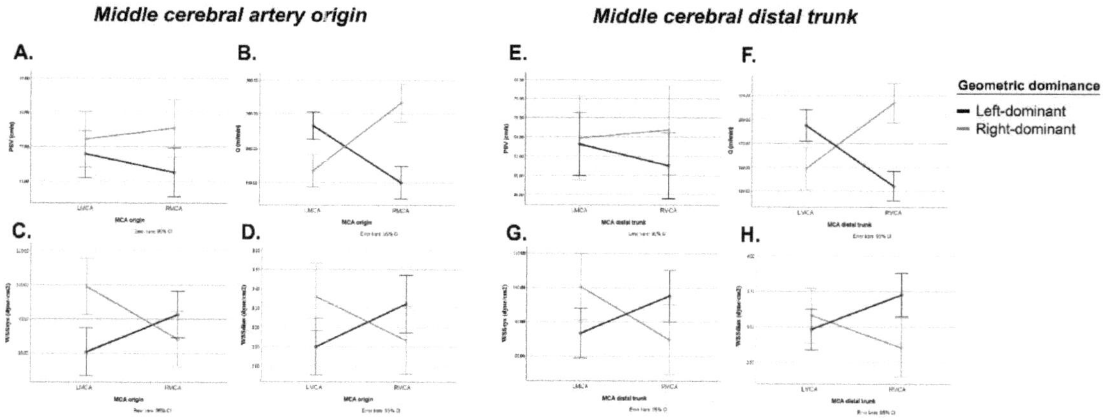

Figure 2. Interactions between haemodynamic and arterial wall shear stress variables and left versus right middle cerebral arteries at the origin (**A–D**) and distal trunk (**E–H**) in left and right dominant neonates. The haemodynamic variables are peak systolic volume (**A**,**E**) and blood flow volume (**B**,**F**). The shear stress variables are systolic wall shear stress (**C**,**G**) and diastolic wall shear stress (**D**,**H**). Abbreviations: LMCA = Left middle cerebral artery; RMCA = Right middle cerebral artery; PSV = Peak systolic velocity; Q = blood flow volume; WSSSYS = systolic wall shear stress; WSSDIAS = diastolic wall shear stress. Error bars show the 95% confidence interval.

The influence of the neonatal arterial geometry on the hemodynamics of these two sites varied, in that the effect of a structural dominance was more pervasive at the origin across most blood flow velocity and flow volume measures (Tables 3 and 4; Figure 2). More specifically, interactions were found in the peak systolic velocity, average velocity, and blood flow volume. Structural dominance of the distal middle cerebral trunk did not significantly influence the arterial velocity (peak systolic, end-diastolic, and mean velocities), but a significant influence of the geometry was seen in the blood flow volume.

The resistance to the blood flow caused by the microvascular bed distal to the site of measurement did not significantly interact with the arterial geometry (Table 4). However, the main effect on the arterial resistance was found at the middle cerebral distal trunk, where the resistance distal to the middle cerebral artery trunk terminus was higher across both cerebral hemispheres in neonates with a rightward geometric dominance ($p = 0.031$). This main effect was also reflected in pulsatility indices.

No lateral differences in the arterial resistance were noted at the middle cerebral artery origin. A lateral difference in the arterial resistance was evident at the distal trunk, where neonates with a leftward structural dominance had a higher resistance and pulsatility index in the left cerebral hemisphere than they had in the right (Table 3). No left–right differences were found in those with a rightward arterial dominance.

3.4. Shearing Stress Differences

The neonatal arterial geometry differentially influenced peak systolic and end diastolic shearing forces at both arterial sites (Table 4; Figure 2). In participants with a leftward structural dominance, the peak systolic and end diastolic shear stresses were significantly higher on the right-side than on the left, and the converse was seen in right-dominant neonates.

4. Discussion

In adults, left–right asymmetries are normal attributes of cerebral perfusion, akin to well-established asymmetries in brain morphology [26,27]. Cerebral arterial diameters and blood flows have been investigated in neonates for a variety of largely clinical ends. Studies of diameters are restricted to autopsy series [28,29]. Blood flow velocity is commonly measured in vivo for routine clinical purposes [30–35].

To our knowledge, this is the first intentional investigation of structurofunctional neonatal cerebral arterial asymmetries in healthy term neonates at rest. Differences in diameter were found at each arterial site of interest, and the corresponding hemodynamics were biased towards larger arterial calibers. Leftward hemodynamic biases were found in neonates with larger arteries in the left cerebral hemisphere (left-dominant), while rightward hemodynamic biases were found in neonates with larger arteries in the right cerebral hemisphere (right-dominant). Very few neonates (<8%) showed an absence of lateral differences in arterial diameter.

The pattern of asymmetry in middle cerebral Doppler waveforms differed between left- and right-dominant groups. Left-dominant neonates were typified by impressive differences in left–right peak systolic velocities that disappeared at the end systole. This peak systolic effect was absent bilaterally in neonates with larger arterial diameters on the right.

Although pulsatility and resistance indices are frequently used in clinical studies, the interpretation of these variables is dependent on several factors such as vascular resistance, arterial compliance, and the driving force of the arterial pulse wave [36]. Structural dominance did not play a role in resistance and pulsatility differences. Arterial pulsatility was not laterally biased in right-dominant neonates, but in left-dominant neonates pulsatility was left biased in the distal trunk of the middle cerebral artery. If one were to apply a traditional interpretation [37] to these findings, the degree of resistance in the cortical microvascular bed distal to the middle cerebral artery would be predicted to be higher in the left hemisphere of most neonates. Higher indices in the left middle cerebral artery would, in turn, indicate a decreased end-diastolic velocity, rendering the left hemisphere more prone to disorders such as stroke or venous infarcts, and left-biased resistance and pulsatility asymmetries in neonates have been documented previously [38].

The arterial endothelial wall shear stress exerts a key influence on the genesis of vascular pathology [39], in that a high shear stress has a protective effect on the endothelium [40]. The pathogenesis of the higher left-than-right incidence of cerebrovascular pathology in adults [41,42] and neonates [43] has been elusive, but clarification might be gained from the overall blood flow and wall shear stress asymmetries reported here.

The present findings show that a systematic leftward arterial bias in wall shear stress is detectible in healthy term neonates. Shearing asymmetries systematically disadvantaged the left hemispheric endothelium with lower left-than-right peak systolic and end-diastolic endothelial shearing forces in neonates with larger left-sided arteries. The converse was seen in right-dominant neonates. This asymmetry therefore increased the neurovascular vulnerability in the left cerebral hemisphere of most healthy term neonates.

The findings of this study are consistent with the hypothesis that the wall shear stress varies according to geometric and behavioral lateralization in the neonatal cerebral arterial trunk. This adds to literature [44] demonstrating that the wall shear stress varies with location across the cardiovascular system. These findings therefore bring Murray's law of constant shearing forces throughout the arterial system [45] into question.

The ontogenesis of atherosclerosis begins very early in life. While incipient atherosclerotic changes are minor in most cases, the process can be accelerated in the presence of a variety of conditions [46]. Menshawi and colleagues [47] postulated that individuals born with an unfavorable arterial geometry are more susceptible to the atherosclerotic effects of traditional vascular risk factors. Although left-lateralized lesions are not inevitable, the predisposing effects of "atherosclerosis-enabling" anatomy reported here might provide the framework for a greater left-than-right incidence of cerebrovascular pathology.

If the present findings are stable across the lifespan and are also consistently discernible in adults, extended exposure to a lateralized arterial vulnerability might also shed light on the ontogenesis of leftward biases in the carotid intima-media wall thickness [48]; plaque incidence, thickness, and instability, as well as large-vessel ischemic events in adults [3].

There are documented associations between handedness and the left arterial intimal wall thickness of the carotids [49], as well as left-handedness and a lower risk of sudden death from brain infarction (typically associated with left-hemispheric stroke [50]. In an adult study, some of the present authors showed that the arterial length, diameter, resistance to blood flow, velocity, and volume flow rate are asymmetric and are intimately related to hand preference and proficiency, raising the possibility that these structurofunctional asymmetries arose in adaptation to greater metabolic demands in the dominant hemisphere in anticipation of the emergence of lateralized cognitive and behavioral functions [51].

Our data show that the asymmetric vascular ground plan found in adults is present in neonates. Ultimately, routine investigations of the neonatal brain should proceed on the expectation that asymmetries in the middle cerebral arteries are a normal attribute of the lateral cortical supply. Ironically, the lateralized neurovascular framework within which language develops might also contain the seeds of its most significant cerebrovascular threat.

Author Contributions: Conceptualization, A.J.v.V., M.S. (Michael Saling) and M.S. (Mark Solms); methodology, A.J.v.V. and S.R.; software, A.J.v.V. and S.R.; validation, S.R.; formal analysis, A.J.v.V.; investigation, A.J.v.V.; resources, S.R. and J.C.; data curation, A.J.v.V.; writing—original draft preparation, A.J.v.V. and M.S. (Michael Saling); writing—review and editing, A.J.v.V., M.S. (Michael Saling), P.A., S.R., M.S. (Mark Solms) and J.C.; visualization, A.J.v.V.; supervision, M.S. (Michael Saling); project administration, A.J.v.V.; funding acquisition, M.S. (Mark Solms). All authors have read and agreed to the published version of the manuscript.

Funding: This research received no external funding.

Institutional Review Board Statement: Royal Women's Hospital Human Research Ethics Committee Project 16/18—Cerebral arterial asymmetries in the neonates.

Informed Consent Statement: Informed consent was obtained from the parents or legal guardians of all neonates involved in the study.

Data Availability Statement: The data presented in this study are available on request from the corresponding author. The data are not publicly available due to privacy reasons.

Acknowledgments: We are indebted the Royal Women's Hospital and Frances Perry House for providing the platform for us to conduct our study. We would also like to thank Julie Archbold for her help in calibrating the Doppler ultrasound machine. The financial assistance of the University of Melbourne is acknowledged.

Conflicts of Interest: The authors declare no conflict of interest.

References

1. Foerch, C.; Misselwitz, B.; Sitzer, M.; Berger, K.; Steinmetz, H.; Neumann-Haefelin, T. Difference in recognition of right and left hemispheric stroke. *Lancet* **2005**, *366*, 392–393. [CrossRef]
2. Hedna, V.S.; Bodhit, A.N.; Ansari, S.; Falchook, A.D.; Stead, L.; Heilman, K.M.; Waters, M.F. Hemispheric Differences in Ischemic Stroke: Is Left-Hemisphere Stroke More Common? *J. Clin. Neurol.* **2013**, *9*, 97–102. [CrossRef] [PubMed]
3. Selwaness, M.; van den Bouwhuijsen, Q.; van Onkelen, R.S.; Hofman, A.; Franco, O.; van der Lugt, A.; Wentzel, J.J.; Vernooij, M. Atherosclerotic plaque in the left carotid artery is more vulnerable than in the right. *Stroke* **2014**, *45*, 3226–3230. [CrossRef] [PubMed]
4. Denarié, N.; Gariepy, J.; Chironi, G.; Massonneau, M.; Laskri, F.; Salomon, J.; Levenson, J.; Simon, A. Distribution of ultrasonographically-assessed dimensions of common carotid arteries in healthy adults of both sexes. *Atherosclerosis* **2000**, *148*, 297–302. [CrossRef]
5. Cagnie, B.; Petrovic, M.; Voet, D.; Barbaix, E.; Cambier, D. Vertebral artery dominance and hand preference: Is there a correlation? *Man. Ther.* **2006**, *11*, 153–156. [CrossRef] [PubMed]
6. Luo, X.; Yang, Y.; Cao, T.; Li, Z. Differences in left and right carotid intima–media thickness and the associated risk factors. *Clin. Radiol.* **2011**, *66*, 393–398. [CrossRef]
7. Lin, P.-Y.; Roche-Labarbe, N.; Dehaes, M.; Fenoglio, A.; Grant, P.E.; Franceschini, M.A. Regional and Hemispheric Asymmetries of Cerebral Hemodynamic and Oxygen Metabolism in Newborns. *Cereb. Cortex* **2012**, *23*, 339–348. [CrossRef]
8. Zhao, M.; Amin-Hanjani, S.; Ruland, S.; Curcio, A.; Ostergren, L.; Charbel, F. Regional Cerebral Blood Flow Using Quantitative MR Angiography. *Am. J. Neuroradiol.* **2007**, *28*, 1470–1473. [CrossRef]
9. Kamath, S. Observations on the length and diameter of vessels forming the circle of Willis. *J. Anat.* **1981**, *133*, 419–423.
10. Baldassarre, D.; Amato, M.; Bondioli, A.; Sirtori, C.R.; Tremoli, E. Carotid artery intima-media thickness measured by ultrasonography in normal clinical practice correlates well with atherosclerosis risk factors. *Stroke* **2000**, *31*, 2426–2430. [CrossRef]
11. Herisson, F.; Heymann, M.-F.; Chétiveaux, M.; Charrier, C.; Battaglia, S.; Pilet, P.; Rouillon, T.; Krempf, M.; Lemarchand, P.; Heymann, D.; et al. Carotid and femoral atherosclerotic plaques show different morphology. *Atherosclerosis* **2011**, *216*, 348–354. [CrossRef] [PubMed]
12. Oxenham, H.; Sharpe, N. Cardiovascular aging and heart failure. *Eur. J. Hear. Fail.* **2003**, *5*, 427–434. [CrossRef]
13. Guzzetta, F.; Shackelford, G.D.; Volpe, S.; Perlman, J.M.; Volpe, J.J. Periventricular intraparenchymal endodensities in the premature newborn: Critical determinant of neurological outcome. *Pediatrics* **1986**, *78*, 995–1006. [CrossRef] [PubMed]
14. Perlman, J.M.; Rollins, N.K.; Evans, D. Neonatal stroke: Clinical characteristics and cerebral blood flow velocity measurements. *Pediatr. Neurol.* **1994**, *11*, 281–284. [CrossRef]
15. Uvebrant, P. Hemiplegic cerebral palsy. Aetiology and outcome. *Acta Paediatr. Scand. Suppl.* **1988**, *345*, 1–100. [CrossRef]
16. Miller, V. Neonatal cerebral infarction. *Semin. Pediatr. Neurol.* **2000**, *7*, 278–288. [CrossRef]
17. Coker, S.B.; Beltran, R.S.; Myers, T.F.; Hmura, L. Neonatal stroke: Description of patients and investigation into pathogenesis. *Pediatr. Neurol.* **1988**, *4*, 219–223. [CrossRef]
18. Dahl, A.; Russell, D.; Nyberg-Hansen, R.; Rootwelt, K. A Comparison of Regional Cerebral Blood Flow and Middle Cerebral Artery Blood Flow Velocities: Simultaneous Measurements in Healthy Subjects. *J. Cereb. Blood Flow Metab.* **1992**, *12*, 1049–1054. [CrossRef]
19. Oktar, S.; Yücel, C.; Karaosmanoglu, D.; Akkan, K.; Ozdemir, H.; Tokgoz, N.; Tali, T. Blood-Flow Volume Quantification in Internal Carotid and Vertebral Arteries: Comparison of 3 Different Ultrasound Techniques with Phase-Contrast MR Imaging. *Am. J. Neuroradiol.* **2006**, *27*, 363–369.
20. Hayashi, T.; Ichiyama, T.; Uchida, M.; Tashiro, N.; Tanaka, H. Evaluation by color Doppler and pulsed Doppler sonography of blood-flow velocities in intracranial-arteries during the early neonatal-period. *Eur. J. Pediatr.* **1922**, *151*, 461–465. [CrossRef]
21. Anwarm, M.A.; Bignall, W.; River, R.P.A. The variation with gestational age of the rheological properties of the blood of the new-born. *Brit. J. Haematol.* **1994**, *86*, 163–168. [CrossRef]
22. Wu, Y.-C.; Hsieh, W.-S.; Hsu, C.-H.; Chiu, N.-C.; Chou, H.-C.; Chen, C.-Y.; Peng, S.-F.; Hung, H.-Y.; Chang, J.-H.; Chen, W.J.; et al. Relationship of Neonatal Cerebral Blood Flow Velocity Asymmetry with Early Motor, Cognitive and Language Development in Term Infants. *Ultrasound Med. Biol.* **2013**, *39*, 797–803. [CrossRef] [PubMed]
23. Jahromi, A.S.; Cinà, C.S.; Liu, Y.; Clase, C.M. Sensitivity and specificity of color duplex ultrasound measurement in the estimation of internal carotid artery stenosis: A systematic review and meta-analysis. *J. Vasc. Surg.* **2005**, *41*, 962–972. [CrossRef] [PubMed]
24. Bishop, C.C.; Powell, S.; Rutt, D.; Browse, N.L. Transcranial Doppler measurement of middle cerebral artery blood flow velocity: A validation study. *Stroke* **1986**, *17*, 913–915. [CrossRef]
25. Kamouchi, M.; Kishikawa, K.; Okada, Y.; Inoue, T.; Ibayashi, S.; Iida, M. Reappraisal of Flow Velocity Ratio in Common Carotid Artery to Predict Hemodynamic Change in Carotid Stenosis. *Am. J. Neuroradiol.* **2005**, *26*, 957–962. [PubMed]

26. Leutin, V.P.; Pystina, E.A.; Yarosh, S.V. Linear Blood Velocity in Arteries of the Brain Hemispheres in Left-Handers and Right-Handers during Hypoxia. *Hum. Physiol.* **2004**, *30*, 290–292. [CrossRef]
27. Willis, M.W.; Ketter, T.A.; Kimbrell, T.A.; George, M.S.; Herscovitch, P.; Danielson, A.L.; Benson, B.E.; Post, R.M. Age, sex and laterality effects on cerebral glucose metabolism in healthy adults. *Psychiatry Res. Neuroimaging* **2002**, *114*, 23–37. [CrossRef]
28. Gielecki, J.; Zurada, A.; Kozlowska, H.; Nowak, D.; Loukas, M. Morphometric and volumetric analysis of middle cerebral artery in human fetus. *Acta Neurobiol. Exp.* **2009**, *69*, 129–137.
29. Seydel, H.G. The diameters of cerebral arteries of the human fetus. *Anat. Rec.* **1964**, *150*, 79–88. [CrossRef]
30. Ebbing, C.; Rasmussen, S.; Kiserud, T. Middle cerebral artery blood flow velocities and pulsatility index and the cerebroplacental pulsatility ratio: Longitudinal reference ranges and terms for serial measurements. *Ultrasound Obstet. Gynecol.* **2007**, *30*, 287–296. [CrossRef]
31. Kehrer, M.; Krägeloh-Mann, I.; Goelz, R.; Schöning, M. The Development of Cerebral Perfusion in Healthy Preterm and Term Neonates. *Neuropediatrics* **2003**, *34*, 281–286. [CrossRef]
32. Pezzati, M.; Dani, C.; Biadaioli, R.; Filippi, L.; Biagiotti, R.; Giani, T.; Rubaltelli, F.F. Early postnatal Doppler assessment of cerebral blood flow velocity in healthy preterm and term infants. *Dev. Med. Child Neurol.* **2002**, *44*, 745–752. [CrossRef]
33. Seffah, J.D.; Swarray-Deen, A. Fetal middle cerebral artery Doppler indices and clinical application at Korle Bu Teaching Hospital, Accra, Ghana. *Int. J. Gynecol. Obstet.* **2016**, *134*, 135–139. [CrossRef]
34. Sinha, A.K.; Cane, C.; Kempley, S.T. Blood flow in the common carotid artery in term and preterm infants: Reproducibility and relation to cardiac output. *Arch. Dis. Child.-Fetal Neonatal Ed.* **2005**, *91*, F31–F35. [CrossRef]
35. Yoshida, H.; Yasuhara, A.; Kobayashi, Y. Transcranial Doppler sonographic studies of cerebral blood flow velocity in neonates. *Pediatr. Neurol.* **1991**, *7*, 105–110. [CrossRef]
36. Bude, R.O.; Rubin, J.M. Relationship between the Resistive Index and Vascular Compliance and Resistance. *Radiology* **1999**, *211*, 411–417. [CrossRef]
37. de Riva, N.; Budohoski, K.P.; Smielewski, P.; Kasprowicz, M.; Zweifel, C.; Steiner, L.A.; Reinhard, M.; Fábregas, N.; Pickard, J.D.; Czosnyka, M. Transcranial Doppler pulsatility index: What it is and what it isn't. *Neurocrit. Care* **2012**, *17*, 58–66. [CrossRef]
38. Ecury-Goossen, G.M.; Raets, M.M.A.; Camfferman, F.A.; Vos, R.H.J.; Van Rosmalen, J.; Reiss, I.K.M.; Govaert, P.; Dudink, J. Resistive indices of cerebral arteries in very preterm infants: Values throughout stay in the neonatal intensive care unit and impact of patent ductus arteriosus. *Pediatr. Radiol.* **2016**, *46*, 1291–1300. [CrossRef]
39. Davies, P.F. Hemodynamic shear stress and the endothelium in cardiovascular pathophysiology. *Nat. Clin. Pr. Cardiovasc. Med.* **2009**, *6*, 16–26. [CrossRef]
40. Slager, C.J.; Wentzel, J.J.; Gijsen, F.J.H.; Schuurbiers, J.C.H.; van der Wal, A.; Van Der Steen, A.F.W.; Serruys, P.W. The role of shear stress in the generation of rupture-prone vulnerable plaques. *Nat. Clin. Pr. Cardiovasc. Med.* **2005**, *2*, 401–407. [CrossRef]
41. Naess, H.; Waje-Andreassen, U.; Thomassen, L.; Myhr, K.M. High incidence of left cerebral infarction among young adults. *J. Stroke Cerebrovasc Dis.* **2006**, *15*, 241–244. [CrossRef]
42. Rodríguez Hernández, S.A.; Kroon, A.A.; van Boxtel, M.P.; Mess, W.H.; Lodder, J.; Jolles, J.; de Leeuw, P.W. Is there a side predilection for cerebrovascular disease? *Hypertension* **2003**, *42*, 56–60. [CrossRef]
43. Lee, J.; Croen, L.A.; Backstrand, K.H.; Yoshida, C.K.; Henning, L.H.; Lindan, C.; Ferriero, D.M.; Fullerton, H.J.; Barkovich, A.J.; Wu, Y.W. Maternal and infant characteristics associated with perinatal arterial stroke in the infant. *JAMA* **2005**, *293*, 723–729. [CrossRef]
44. Dammers, R.; Stifft, F.; Tordoir, J.H.M.; Hameleers, J.M.M.; Hoeks, A.P.G.; Kitslaar, P.J.E.H.M. Shear stress depends on vascular territory: Comparison between common carotid and brachial artery. *J. Appl. Physiol.* **2003**, *94*, 485–489. [CrossRef]
45. Murray, C.D. The Physiological Principle of Minimum Work: I. The Vascular System and the Cost of Blood Volume. *Proc. Natl. Acad. Sci. USA* **1926**, *12*, 207–214. [CrossRef]
46. Hong, Y.M. Atherosclerotic Cardiovascular Disease Beginning in Childhood. *Korean Circ. J.* **2010**, *40*, 1–9. [CrossRef]
47. Menshawy, K.; Mohr, J.P.; Gutierrez, J.G.A. A Functional Perspective on the Embryology and Anatomy of the Cerebral Blood Supply. *J. Stroke* **2015**, *17*, 144–158. [CrossRef]
48. Lemme, C.; Jogestrand, T.; de Faire, U. Carotid intima-media thickness and plaque in borderline hypertension. *Stroke* **1995**, *26*, 34–39.
49. Önbaş, O.; Dane, Ş.; Kantarci, M.; Koplay, M.; Alper, F.; Okur, A. Clinical importance of asymmetry and handedness differences in common carotid artery intima-media thickness. *Int. J. Neurosci.* **2007**, *117*, 433–441. [CrossRef]
50. Algra, A.; Gates, P.C.; Fox, A.J.; Hachinski, V.; Barnett, H.J. North American Symptomatic Carotid Endarterectomy Trial Group: Side of brain infarction and long-term risk of sudden death in patients with symptomatic carotid disease. *Stroke* **2003**, *34*, 2871–2875. [CrossRef]
51. Jansen van Vuuren, A.; Saling, M.M.; Ameen, O.; Naidoo, N.; Solms, M. Hand preference is selectively related to common and internal carotid arterial asymmetry. *Laterality* **2016**, *4*, 377–398. [CrossRef]

symmetry

Review

The Neurological Asymmetry of Self-Face Recognition

Aleksandra Janowska, Brianna Balugas, Matthew Pardillo, Victoria Mistretta, Katherine Chavarria, Janet Brenya, Taylor Shelansky, Vanessa Martinez, Kitty Pagano, Nathira Ahmad, Samantha Zorns, Abigail Straus, Sarah Sierra and Julian Paul Keenan *

The Cognitive Neuroimaging Laboratory, Montclair State University, 207 Science Hall, Montclair, NJ 07043, USA; janowskaa1@montclair.edu (A.J.); balugasb1@mail.montclair.edu (B.B.); pardillom1@montclair.edu (M.P.); 220731@pcti.mobi (V.M.); chavarriak2@mail.montclair.edu (K.C.); brenyaj1@mail.montclair.edu (J.B.); shelanskeyt1@montclair.edu (T.S.); martinezv8@montclair.edu (V.M.); paganok1@mail.montclair.edu (K.P.); ahmadn3@mail.montclair.edu (N.A.); zornss1@montclair.edu (S.Z.); astra20030@sufsdny.org (A.S.); sierras5@montclair.edu (S.S.)

* Correspondence: keenanj@montclair.edu

Abstract: While the desire to uncover the neural correlates of consciousness has taken numerous directions, self-face recognition has been a constant in attempts to isolate aspects of self-awareness. The neuroimaging revolution of the 1990s brought about systematic attempts to isolate the underlying neural basis of self-face recognition. These studies, including some of the first fMRI (functional magnetic resonance imaging) examinations, revealed a right-hemisphere bias for self-face recognition in a diverse set of regions including the insula, the dorsal frontal lobe, the temporal parietal junction, and the medial temporal cortex. In this systematic review, we provide confirmation of these data (which are correlational) which were provided by TMS (transcranial magnetic stimulation) and patients in which direct inhibition or ablation of right-hemisphere regions leads to a disruption or absence of self-face recognition. These data are consistent with a number of theories including a right-hemisphere dominance for self-awareness and/or a right-hemisphere specialization for identifying significant social relationships, including to oneself.

Keywords: symmetry; self-face recognition; right hemisphere; self-awareness

Citation: Janowska, A.; Balugas, B.; Pardillo, M.; Mistretta, V.; Chavarria, K.; Brenya, J.; Shelansky, T.; Martinez, V.; Pagano, K.; Ahmad, N.; et al. The Neurological Asymmetry of Self-Face Recognition. *Symmetry* **2021**, *13*, 1135. https://doi.org/10.3390/sym13071135

Academic Editors: Sergei D. Odintsov and Chiara Spironelli

Received: 1 April 2021
Accepted: 21 June 2021
Published: 25 June 2021

1. Introduction

The evolution of animal nervous system symmetry is complex, with many resulting variants [1–5]. Allowing for numerous phenotypic advantages, including those at both an individual and social/interactive level [6–9], the nervous systems of bilateral organisms have exploited the benefits of a lateralized nervous system for hundreds of millions of years [2,10] (but see [11]).

The human brain is no exception [2,12]. The first impression of the human brain was noted as far back as the Ancient Greeks as two distinct hemispheres. Except in rare cases of severe abnormal development, a human at any stage post-second trimester will anatomically have two distinctly visible hemispheres. These anatomical differences have given rise to functional differences, scientifically noted by Broca, Wernicke, and others in the late 1800s and early 1900s [13]. While language remains the most well-known of human brain lateralization, many other functions appear distinctly prominent in one hemisphere [14–18].

That being said, research concerning left and right hemisphere differences (LH/RH) in the brain appears to trend from 'too simplistic' to 'explains everything'. While Roger Sperry's Nobel Prize in 1981 seemed to cement the legitimacy of exploring hemispheric differences [19], the popular press has run with mythical notions such as people being 'right- or left-brained'. The more measured approach is understanding both the ultimate and proximate reasons for asymmetries.

For example, the motor asymmetries observed in humans and other primates [12,20–22] have led to numerous theories the most useful of which include both the mechanisms and the underlying cost/benefit analyses. An example of this is that human cradling (mother/infant) is performed employing the left arm the majority of the time [8,23–25]. Such a bias is explained by socio-affective communication being RH dominant, which clearly taps into evolutionary explanations (i.e., facilitating social bonding) while also explaining functional brain hemispheric differences. That is, these data help to resolve why certain aspects of emotional communication may be RH dominant and how right-handedness itself may have evolved.

Here we describe the evidence that self-face recognition (SFR) is RH dominant and speculate that it is related to the underlying construct of self-awareness (SA; [26]), and we provide a review of the literature. It is recognized that while the evidence for RH dominance is robust [27], thus far the data are suggestive in terms of a link between SFR and SA [28–30]. In terms of evolution, we know even less. For example, while initial evidence indicates hemispheric differences in terms of SFR in chimpanzees [31,32], other animals (particularly the magpie, who may or may not have SFR: [33,34] may have an entirely different underlying neural structure that supports SFR and, potentially, SA. We therefore are unable to know at this point whether SFR has evolved independently or in a more homologous manner [35].

The cone of uncertainty widens in terms of SFR as we move phenotypically and phylogenetically further from *Homo sapiens*. In humans, we propose SFR as the benchmark against which all measures of self-awareness should be tested. We suggest, in fact, that despite numerous challenges, SFR in the great apes indicates SA, and we predict that in the near future, homologous neural underpinnings in the RH will be discovered that sustain SFR. We equally suggest that as we move to cetaceans, corvids, elephants, etc., SFR becomes more unstable, and both the underlying structures and the underlying indices of SFR become much less clear. Our purpose here is to highlight the main research that underlies these claims. While not a thorough review, this summary is intended to provide a clear and concise overview of SFR.

2. The History of Self-Face Recognition: Measuring Self-Awareness

Questions concerning self-awareness have been posed by almost all humans, including scientists, for millennia. From the Ancient Greek scholars, all the way through Gallup's mirror self-recognition tests in the 1970s, to today's modern brain imaging techniques, self-awareness has always been an intriguing topic to study and investigate.

Greek philosophers, including Socrates (b. 470 B.C.E.), believed that introspection was necessary for humans to be truly cognizant and pure. Plato (b. ca. 428 B.C.E.) took this concept of self even further and stated that introspection was a human obligation and that knowledge of "good" and "self" were needed in order to be honorable and principled. Importantly, Aristotle (b. 384 B.C.E.) took a comparative approach to look at differences between self-awareness in humans and nonhuman animals through studying cognitive intelligence. He concluded that both humans and animals had basic functions, such as sight, smell, taste, etc., but *pure* intellect was only found in humans, which made a large distinction between humans and animals. Additionally, Aristotle was one of the first to attempt to create a relationship between the self, soul, and body [36].

Most famously, the French mathematician René Descartes (b. 1596) took the study of consciousness further, as he is often considered among the first neuroscientists that attempted to localize the self. Many of his ideas are still commonly spoken of today, such as "*Cogito, ergo sum*" ("I think; therefore, I am"), which speculated that the self can exist because it can think of its own existence. Outside of just defining the self, Descartes attempted to actually locate the self in the brain. Although his determination of self in the brain as being located in the pineal gland (due to its centralized position in the brain) was ultimately wrong, his comparative look at humans and animals had a lasting influence [37–40]. Unlike Aristotle, Descartes believed that animals are intelligent but do not have a soul or self. He found that animals do not use language, behave on impulses, and are not adaptable, so they cannot have a self.

The Ancient Greek philosophers and Descartes laid the groundwork for formalizing scientific investigations of the self under the umbrella of psychology. Most famously, Sigmund Freud explained the deep, buried unconscious mind by using the self while simultaneously explaining the self [36]. Carl Jung believed that there were many common selves that people shared to some degree. Jean Piaget believed that children refine their self through assimilation and childhood experiences, which, in turn, play a key role in growing into adulthood. Many philosophers (including Locke, Sartre, Hegel, and Hume) and psychologists (e.g., Seligman, Beck, and Kohler) examined relationships between self-awareness and cognition.

It is noted that many individuals were addressing self/other distinctions as other sciences came into their own. In the 18th century, Carolus Linnaeus, a Swedish botanist, created a binomial classification system for living organisms. While it is known that many of his taxonomic names are still used today, such as kingdom, order, species, etc., we note here that one important classification was grouping humans along with other animals such as monkeys, apes, and bats as primates. Linnaeus remembered a Latin inscription, "*Nosce te ipsum*" ("Know thyself"), translated from the Greek above the Temple of Apollo at Delphi, from which he assumed that the distinction between humans and other primates is the capability for self-recognition and self-knowledge. Therefore, he categorized humans into *Homo sapiens*, wise men, since he believed that self-awareness was the highest form of uniquely human intelligence [41,42].

Up until this time, the self was quite an abstract concept with little concrete evidence, but that slowly changed when the mirror was seen as a tool to be used to measure cognitive abilities and self-awareness. Grant, in 1828, was the first person to our knowledge to use a mirror for a self-recognition study. The study found that monkeys, in general, had a surprised reaction when looking at mirror glass, but orangutans, in particular, had no emotional response to looking at the glass. The exact reaction of the orangutans was not recorded, just the lack of reaction, which was unfortunate, but enough of a reaction to record considering the monkeys' strong reaction to the mirror. Soon after, Charles Darwin was one of the first to suggest using mirror recognition as a measure of higher cognitive abilities. His first recorded mirror test in 1840 examined the behavior of orangutans presented with a mirror, in which he recorded, and later published in 1876, that the orangutans would look at the mirror as if they were seeing another animal [43]. This led to Darwin's conclusion that self-recognition was not an ability of nonhuman animals. Additionally, Darwin studied his 10 children as they grew up, starting in 1839, and did mirror tests with them. His conclusions were that self-awareness and self-knowing were tied to the ability to self-recognize [44].

In 1878, Maximillian Schmidt reported similar findings to Grant and Darwin (that orangutans did not have a self-reaction to a mirror but seemed to understand the reflective properties of the mirror). Schmidt noted that the orangutan was able to identify a human reflection in the mirror of someone standing nearby [45]. These findings were the norm in many studies performed around this time with nonhuman mirror tests. Another study performed by J. von Fischer in 1876 observed monkeys and baboons in front of a mirror and, once again, a negative mirror self-recognition result was reported. It is noted that in these studies, rigorous methods to determine SFR were not employed. The general sense is that an organism was placed in front of a mirror and behavior was observed, such as attacking the mirror or apathy. The main conclusion was that the various monkeys tested by von Fischer reacted to the mirror as if it was another, novel monkey.

Not too long after, in 1889, Wilhelm Preyer, a German researcher, was able to define a definite sign of self through the use of mirrors. He studied how using only language would be an inadequate use to describe the self in children due to the lack of vocabulary, not the lack of understanding of the "I", or ego. He created developmental timelines using mirror recognition, language, and other time measures, such as own-shadow recognition, to pinpoint a child's timeline of self-recognition. Through this, Preyer was able to confirm the use of mirrors for self-recognition tests due to his orderly, thorough reports. Although Preyer did not specifically work with apes, he did record nonhuman animal reactions compared to the human reactions [46]. Unfortunately, most mirror self-recognition researchers did not communicate together at the time, so many went unnoticed, causing a dip in the study of mirror tests for self-recognition.

Outside the occasional mirror test on monkeys, orangutans, and chimpanzees, most mirror self-recognition slowed down, until 1929. Robert and Ada Yerkes found results suggesting no self-interest in mirrors by the nonhuman primates, noting that the animals were seeing 'another animal' in the mirror [47]. Following this, mirror tests were performed rarely and without any true lineage of experiments to follow. In 1940, C. W. Huntley also performed a little-known experiment that recorded human participants having a large emotional reaction to the realization that the recorded voice played back, hand pictures, and handwriting were indeed their own [48]. Arnold Gesell, a Yale child developmentalist, studied similar theories in the early to mid-20th century as Piaget and Preyer and studied many self-indicators of how a child's timeline develops these indicators [49]. Unfortunately, lack of expansion on these previous theories with mirror tests led Gesell's reports to fall through the cracks. In the 1940s, Jacques Lacan suggested that the formation of the self and mirror recognition were correlated, but due to a lack of mirror experimentation, his research was overlooked. In 1954, in a paper for the journal *Human Biology*, a photograph of a chimpanzee named Vicki using a mirror to guide pliers over her teeth was taken, but the authors, Keith and Catherine Hayes, did not discuss its relevance (that there was evidence that a *nonhuman had self-recognition*) and it went unrecognized [50].

3. Gallup's Mirror Self-Recognition Test

The lull of interest in mirror tests finally changed abruptly in 1970 when Gallup published research of a nonhuman-animal positive mirror self-recognition test. Gallup's mirror and mark test combined many previous ideas about self-recognition, and in the article, he commented that there is likely a connection between self-recognition and self-knowing. By creating a test that measured a real physical trait along with a well-thought-out process that eliminated random chance, Gallup made a solid argument for self-recognition in chimpanzees. However, he and others immediately picked up the notion that self-recognition in a mirror may, in fact, be evidence that humans are not the only self-aware organism on this planet.

The first test consisted of a 10-day period of a mirror placed in the testing site with the chimpanzees to allow them to acclimate with the mirror. This created a baseline behavior of the chimpanzees, which was typical mirror behavior, as seen with previous tests where there was no significant reaction to the mirror. Then, the chimpanzees were placed under anesthesia and a small mark that was dry and odorless was placed above the eyebrow where the chimpanzees could not see directly. After waking up from anesthesia, the chimpanzees' behavior was observed without a mirror. It was found that the chimpanzees did not react to their new mark (in the absence of a mirror), no smelling or touching occurred, and therefore, it was concluded that they were unaware of the mark. After this short baseline period, mirrors were placed in the chimpanzee test site again and their reactions were observed. Upon seeing their reflections in the mirror, the chimpanzees would touch and smell the mark and investigate their hands after touching the mark. This indicated that the animals recognized that the mark had not been there previously and that they had to use a mirror to find the mark. This test itself was a breakthrough, but Gallup conducted two more tests to confirm this conclusion.

The second test was almost identical to the first, with the exception of the initial 10-day pre-mark mirror exposure. Gallup hypothesized that without the previous mirror exposure, the chimpanzees would not react to the mark as being odd on their face, which proved to be true. The chimpanzees were given the mark and indeed did not react to it in any significant manner, leading Gallup to conclude that the first group achieved mirror self-recognition (MSR). The third test Gallup conducted was identical to the first, but instead of chimpanzees, he used monkeys. He observed their reactions pre-mark and post-mark with a mirror, and the monkeys gave no sign of MSR [51].

The three tests put together create a clear, concise result that chimpanzees are able to recognize themselves in the mirror. Gallup published his findings in 1970 not only discussing the test and the breakthrough results he found, but also contemplating the higher consciousness of nonhuman animals. Before then, it was assumed that humans have the highest form of intelligence and cognition and that animals have some intelligence but not necessarily a "soul" or self. With these findings, Gallup opened up the discussion of nonhuman animals possibly having a higher-order cognitive process and internal world. With this mirror and mark test, Gallup also sparked a new method of research with the ability to test consciousness hypotheses.

Since Gallup's initial findings, the literature has swayed from conservative (only humans, chimpanzees, and orangutans have self-face recognition) to today with more liberal interpretations that include the addition of animals such as elephants, dolphins, and magpies [52].

4. SFR in Animals

The advantages of Gallup's test are many. It does not require language, which means it can be used in pre-/non-linguistic humans and nonhumans. It requires no special equipment; the equipment needed is portable and can be used in any environment. It requires little training to conduct, it is inexpensive, and it does not take months to perform. The test is noninvasive and generally culture-fair, though there are cultures where mirror exposure may be limited [42].

Gallup's test involves a number of aspects that present difficulties in testing both human and nonhuman populations. The test relies on vision and memory, as well as intact motor systems, and alterations to these can influence results. The test requires the individual to inspect the mark to 'pass', which in the original form meant having limbs/arms capable of touching the mark. The test requires some aspect of 'caring' that one's image is altered to the point of touching the mark. This last point is a major issue when testing special groups of humans, such as those with autism, who may know that there is a mark but not feel compelled to inspect the mark [53,54]. While there are other issues (e.g., attentional processing, what qualifies as a successful touch, and the role of training vs. spontaneous SFR), it should be apparent that direct comparison across populations is difficult if not impossible.

For example, dolphins and other marine mammals are notorious for being a 'challenging' population to test as mark-directed responses are near impossible to conduct [55–57]. Self-recognition has been claimed not via mark-directed touching but via inspection of the mark in a mirror and self-directed behavior conducted by use of the mirror. Horses have been tested via 'scraping' behavior of a mark on the cheek [58]. Elephants are considered to pass if they make a trunk-directed response [59–61].

It is possible that tests for SFR are most opaque in terms of the avians [62–66]. While an early report indicated magpies passing a modified version of Gallup's test [64], this finding has not been consistent [33]. B.F. Skinner was one of the first to perform SFR avian studies when he and his colleagues conditioned pigeons to peck at a spot [67]; see [65,66], which caused others to emphasize that SFR must be spontaneous and not 'taught' [68]. Other attempts to modify the test have included making the test more phenotypically relevant by using olfaction rather than vision [69]. While this is but a brief summary, we remain skeptical about SFR in non-great apes (with the exception of dolphins) and conclude that more evidence is needed before we consider organisms outside of apes and dolphins to have SFR abilities [52].

5. Nonhumans and Brain Symmetry

Before presenting evidence of a lateralized system for SFR, it is worth examining the sparse evidence of the neural correlates of SFR in nonhumans. Bill Hopkins and his team examined mirror self-recognizers (MSR+) vs. non-recognizers (MSR−). Employing diffusion tensor imaging, a lateralized MSR+ vs. MSR− difference was found in the superior longitudinal fasciculus (including frontal lobe areas). The authors indicated that SFR was associated with a greater RH asymmetry [31].

A follow-up study found that MSR+ chimpanzees had increased cortical thickness bilaterally in the caudal anterior cingulate gyrus (mostly in the right hemisphere) and thinner cortex in the central portion of the pre- and postcentral gyri, primarily in the left hemisphere [32].

These results are too limited to draw distinct evolutionary patterns, but they are consistent with the RH asymmetry observed in humans in terms of SFR. That is, it is unclear whether the neural architecture that provides MSR in nonhumans is homologous to what occurs in humans and less clear whether there is a direct evolutionary path. However, taken together with what we find in humans, the data suggest that at least in chimpanzees, there are homologies rather than analogies. That is, the early data indicate that the same rightward bias for MSR that exists in chimpanzees also exists in humans.

6. Functional Imaging Indicates Right-Hemisphere Dominance in Self-Face Recognition

In the early 1990s through to today, research has examined how the brain actually allows MSR, treating it as an exceptional ability. Before neuroimaging, however, there were indicators of a possible right-hemisphere (RH) bias, as disorders of RH neural structures sometimes lead to a lack of own-body recognition [41,70–73] and disorders with self-awareness deficits appeared to be similar to RH disorder. Early attempts to determine the correlates of self-face recognition were made by pioneers such as Preilowski [74], who was the first to suggest a RH bias even though his methods involved indirect indicators.

In an early attempt to test Preilowski's hypotheses, we employed the WADA method in which the anterior portion of one hemisphere is anesthetized. Using self-face morphs (e.g., self-morphed to Marylin Monroe), it was found that following RH anesthesia, patients had significant difficulty recognizing their own face (Figure 1 [75]). There were early attempts to use lateralized hand response differences as a further test of RH SFR [65,66]. However, as is the case in much of cognitive neuroscience, fMRI (functional magnetic resonance imaging) dominates the literature.

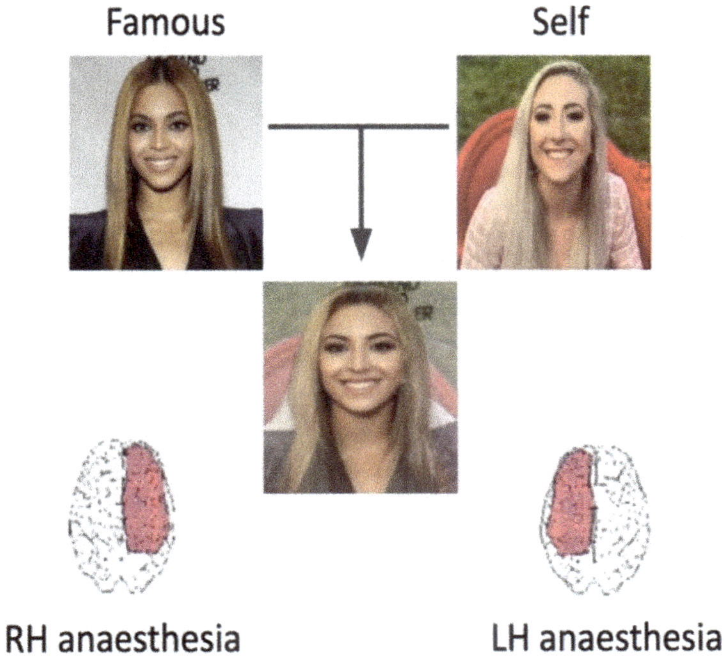

Figure 1. Numerous methods have been employed to reach the conclusion that there is a RH lateralization in terms of MSR/SFR. Anesthesia applied to the brain in either the right or left hemisphere leads to differences in SFR. Namely, patients without a fully functioning RH see morphed images as not themselves. In this case, they report the image as Beyoncé under RH anesthesia conditions and their own face under LH anesthesia conditions [75].

Functional MRI (fMRI) of the self-face commenced in the mid-1990s with simple designs involving either unaltered self-faces contrasting with other faces or basic morphs [68]. Over time, the designs became more sophisticated, including examining affect, psychiatric disorders [76–80], and family [81]. While there is an overwhelming bias for RH activity in these studies, fMRI also revealed both a wide distribution of regions and the notion that many variables influence how the brain processes one's own face [82,83]. Much of the variability across the studies is for reasons unknown as the studies have not been replicated. Therefore, the fact that in one study a face is presented for a certain duration (for example) may be the factor, or it may be the task itself. We do find convincing data that suggest the more the SFR task engages self-reflection, the greater the bilateral medial frontal activity is [27].

Sugiura's group was one of the pioneers in determining the cortical correlates of SFR using fMRI [84–91]. Importantly, it was found that the brain has a number of distinct regions/networks associated with self–other distinctions. While there is overlap, the critical finding is that RH activation works in concert with medial frontal areas when SFR is performed in a social context. In other words, the social component of SFR appears to draw on the RH, as well as medial frontal networks.

Morita and colleagues also conducted numerous SFR fMRI studies and found a consistent RH bias [78,92–96]. In terms of brain symmetry, she solved a problem that has baffled researchers—the role that handedness plays in the lateralization of SFR [92]. They discovered that most right-handed participants exhibited a RH SFR bias. Likewise, most left-handers had RH bias, but there were significant numbers that had LH-localized SFR comparted to the right-handers. Therefore, it appears that, like handedness itself, which sometimes involves a shift in verbal language dominance, SFR may also shift to the LH in concert with some cases of left-handedness.

It may be surprising that three meta-analyses have been performed on SFR and the brain [27,97,98]. All three indicate a RH bias in SFR, though there are bilateral activations at a much lesser level. A key question that fMRI has helped to answer is the relationship between SFR and metacognition, either of oneself (SA) or of others (Theory of Mind: TOM). That is, if there is a relationship between SFR and TOM at a neurological level, such a relationship may be not just spurious but related in a meaningful way behaviorally and at an evolutionary level. It was found in two of the meta-analyses that SFR activates RH networks and overlaps with cortical midline structures in terms of metacognition, most likely critically involving medial regions of the frontal lobe [27,98]. We therefore conclude, based on fMRI, that the RH is likely necessary but not sufficient for SA.

Functional MRI provides only a correlational relationship between brain activity and behavior. By the 2000s, research was regularly appearing demonstrating a causal relationship. That is, by disrupting regions of the RH using noninvasive techniques, researchers were discovering that SFR was not just correlated with the RH but was actually involved in a causal relationship. The main methods employed thus far involve a version of a 'virtual lesion' in which a brain area is either temporarily taken offline or temporarily severely inhibited [99,100]. Basically, different regions of the brain were disrupted, and the subsequent changes in self-face recognition were measured, which established causality.

While early attempts using TMS (transcranial magnetic stimulation) did distinguish the right hemisphere as being necessary for self-face recognition [75], the most elegant of these studies was provided by Lucina Uddin and her colleagues [101]. She found that right parietal TMS disrupted self-face recognition, whereas left parietal TMS did not. Working in a somewhat 'backwards' manner, Uddin took these causal data and supported them (i.e., RH dominance) with correlational imaging data [102,103].

More recent studies have found that disruption of the RH is more dramatic for those individuals with subclinical grandiose narcissism [104]. That is, as narcissistic traits increase, TMS delivered to the RH causes a greater disruption of recognition of one's own face. This study, unlike studies of autism (see below), is one in which we see an excess of SA (rather than a deficit) correlating with some measure of SFR. We speculate that hyperactivity of self-associated neural networks associated with narcissistic traits leads to a steeper decline in function when disrupted (compared to normal activity), though this hypothesis needs further testing.

An even more interesting study revealed that the RH bias for self-faces may in fact be subconscious, below one even needing to identify whether the face is their own. Using a mental rotation task involving either one's own face or the face of another, Zeugin and his colleagues found that RH parietal TMS disrupted mental rotation of self-face compared to familiar faces in general [105]. This might indicate that the 'specialness' of the self-face is much 'deeper' in one's cognitive schema and does not rely on conscious representation. It would be interesting to test narcissistic traits (as in [104]) to see whether there are similar contributions of narcissism at the implicit level, as we previously observed at the explicit level.

Confirming these TMS studies in terms of brain/behavior disruption, a similar technique known as tDCS (transcranial direct cortical stimulation) was employed to alter the brain with self-face identification being measured. It was found that disruption of the temporo-parietal lobe in the RH disrupted self-face perception [106].

As is often the case in neuroscience, we have to ask whether we can extrapolate the data from the lab to the clinic to the 'real world'. While patient data from post-callosotomy (split-brain) surgery individuals [103] to those with autism [53,78,103] suggested a possible RH bias, a more specific method for detailing potential asymmetries is need. Specifically, we need to turn to those that lose self-face recognition in the absence of prosopagnosia (i.e., general face-recognition loss).

7. Patient Data: Delusional Misidentification Syndrome

While TMS and other neuroimaging points towards a RH bias in SFR, the data become infinitely stronger if matched within a patient population. Losing the ability to recognize the self is a hallmark of late-stage dementia. However, this deficit tells us little about the locality of SFR. Although rare, there are in fact cases where the loss of SFR occurs against the background of relatively stable cognitive processing [107].

Delusional misidentification syndrome (DMS) refers to circumstances in which patients form a fixed, distorted belief regarding the identity of a person, place, or object [107,108]. These disorders include Capgras syndrome, in which a person that was once very familiar to the patient (e.g., their husband) is now perceived as a stranger. Amazingly, there are cases where Capgras exists *exclusively* for the self, in which a patient misidentifies him/herself as being either a stranger they have not met or a different, familiar person that is not the patient. This disorder is rare, and there are only a few cases in the existing literature. It is not agreed upon yet what the naming of this disorder is, but it tends to be referred to as mirror sign or Capgras for the self [109–114].

In the most substantive review to this point, David Roane et al. [115] examined 24 case reports of the mirror sign. Of note is that within most of these cases, the patients were successful in correctly identifying the mirror images of *others*, signifying that they do not have a general impairment in recognition of familiar faces and they understand what a mirror does [116]. That is, the loss of face recognition was exclusive to the self and not due to prosopagnosia or a lack of understanding of what a mirror does/how it functions. In terms of localization, there was a diverse range of methods employed to obtain anatomical and functional data. Of the 24 patients, 9 of them had clear evidence for RH damage including the "parietal, temporoparietal, occipito-temporal, dorsolateral frontal cortex, basal ganglia, and thalamus" [115]. Out of 24 cases detailing mirror sign, imaging data were reported in 20 of them. Of the 18 MRI findings reported, 13 showed patients with mild generalized atrophy, as well as atrophy in specific regions within the right hemisphere of the brain. Further, PET and EEG findings supported RH dysfunction, displaying hypermetabolism in the right prefrontal, parietal, and occipital–temporal cortex and right temporal slowing. Thus, DMS can be considered a RH disorder [70–73].

In particular, a case report of the mirror image followed an elderly 77-year-old right-handed woman by the name of SP. SP was hearing impaired from a young age and was known to have communicated through sign language and also by lip-reading those around her. The patient's misidentification was regarded as highly selective as she was capable of readily identifying others in the mirror, though she regarded her own reflection as "the other SP", a companion of sorts. As expected, SP's lack of self-face recognition was supported upon her neurological examination; her MRI scan demonstrated clear RH damage [41].

The authors concluded that the association of RH involvement is consistent with previous work linking self-recognition to the right prefrontal and right frontoparietal cortex [42]. Overall, the most common findings were localized to the RH, which is not an unexpected finding [115].

8. Why Does This Make Sense?

The question remains as to why SFR is lateralized to the RH, and the clearest answer is that we do not know for certain. Most suggest that the RH has a specialization for social processing [117–122], though this ability is not exclusive to the RH. Rather, the RH (specifically the right TPJ) appears to be critical in correctly making self–other distinctions—an ability needed for empathy and TOM [123,124]. In fact, it is now well established that the right TPJ is critical for 'feeling another's pain' and the ability to apply one's own feelings to another [123].

Overall, social patterns and self–other distinctions are biased to the RH, and it is plausible that SFR is tapping into this to the point that in the absence of SFR, there may be social deficits. For example, there are SFR deficits in pervasive developmental disorders [53,54,80,125–130], though the range of deficit is from severe to nondetected. Quevedo's group furthered this discussion by examining SFR in clinical populations including those with depression. Persons rated high in suicidal affect and cognitions, for example, have a unique neural response to the self-face compared to those without; self-faces involve differential neural circuits including the amygdala depending on clinical diagnosis and symptomatology [131–135]. These data indicate that SFR in its absence may indicate a lack of SA (i.e., autism) and that SFR in its presence may indicate different degrees of SA in different populations. In anorexia nervosa, a condition with a lack of accurate SA, there is difficulty in SFR [136]. Insight into one's own schizophrenia (i.e., schizophrenics that have awareness of their condition) correlates positively with SFR [137]. Further, disorders of consciousness also correlate (as indicated by physiological measures) with SFR [137].

The false belief tasks that are prevalent in TOM testing tap into the notion that the human brain is capable of modeling two brains/belief systems simultaneously [138–140]. This differentiation is critical for our social interactions as humans. For example, I need not be in pain to know that you are in pain. Parenting, reproducing, successful predation and predator avoidance, etc., are all enhanced by the ability to separate one's own thoughts from another's thoughts. Thus, and somewhat ironically, empathy and TOM are increased by the ability to separate one's SA from another's SA.

We propose that RH development and laterality has evolved for numerous reasons, though like others, TOM is likely the main impetus. What we propose, however, is that the role of teaching is overlooked in this domain. Humans are specialists at active teaching, and given the plasticity of the human brain, active teaching is a critical component of human survival. We examined the role of TOM in teaching and learning, as well as teaching communication, in a task that involved building simple Lego models and found that the best teachers have high TOM [141]. This task was demanding in terms of active teaching and social interaction, and of interest is that the learner's TOM was not critical (i.e., a learner can have low TOM and still learn), which models human–infant interactions.

Taken together, SFR appears to be an indicator of SA, and as SA fluctuates, so does SFR. RH patients are prone to deficits of SA [71–73,115,142]; thus, it is not surprising to see that SFR, SA, and the RH are related. It is worth noting that we do not know whether other non-ape organisms achieve SFR, or even if SFR is possible with very different brain mechanisms or if one needs a highly lateralized brain to have SFR and/or SA [42]. Finally, we are still unclear as to whether SA is truly indicated by SFR. As it is, 30 years of neuroimaging and patient data have left us with more questions than answers.

Author Contributions: Conceptualization, A.J., B.B., M.P., K.C., J.B., N.A., S.Z. and J.P.K.; methodology; software, N.A.; validation N.A.; formal analysis, N.A.; investigation, A.J., B.B., M.P., V.M. (Victoria Mistretta), K.C., J.B., T.S. and J.P.K.; resources, N.A.; data curation, N.A.; writing—original draft preparation, A.J., B.B., M.P., V.M. (Vanessa Martinez), K.C., J.B., T.S. and J.P.K.; writing—review and editing, A.J., B.B., M.P., V.M. (Victoria Mistretta), K.C., J.B., T.S., V.M. (Vanessa Martinez), K.P., N.A., S.Z., A.S., S.S. and J.P.K.; visualization, A.J. and J.P.K.; supervision, J.P.K.; project administration, J.P.K.; funding acquisition, J.P.K., K.C. and J.B. All authors have read and agreed to the published version of the manuscript.

Funding: The work was funded by LSAMP (Louis Stokes Alliance for Minority Participation), The Crawford Foundation, and the Wehner Fund. Josh and Judy Weston provided funding as well as the Kennedy Foundation.

Institutional Review Board Statement: Not applicable.

Informed Consent Statement: Not applicable.

Data Availability Statement: Not applicable.

Conflicts of Interest: The authors declare no conflict of interest.

References

1. Aboitiz, F.; Montiel, J.; Morales, D.; Concha, M. Evolutionary divergence of the reptilian and the mammalian brains: Considerations on connectivity and development. *Brain Res. Rev.* **2002**, *39*, 141–153. [CrossRef]
2. Corballis, M.C. Of mice and men—And lopsided birds. *Cortex* **2008**, *44*, 3–7. [CrossRef] [PubMed]
3. Corballis, M.C. Evolution of cerebral asymmetry. *Chang. Brains Appl. Brain Plast. Adv. Recover Hum. Abil.* **2019**, *250*, 153–178. [CrossRef]
4. Genikhovich, G.; Technau, U. On the evolution of bilaterality. *Development* **2017**, *144*, 3392–3404. [CrossRef] [PubMed]
5. Manuel, M. Early evolution of symmetry and polarity in metazoan body plans. *C. R. Biol.* **2009**, *332*, 184–209. [CrossRef]
6. Rogers, L.J.; Vallortigara, G. Complementary Specializations of the Left and Right Sides of the Honeybee Brain. *Front. Psychol.* **2019**, *10*, 280. [CrossRef]
7. Rogers, L.J.; Zucca, P.; Vallortigara, G. Advantages of having a lateralized brain. *Proc. R. Soc. B Boil. Sci.* **2004**, *271*, S420–S422. [CrossRef]
8. Tommasi, L. Mechanisms and functions of brain and behavioural asymmetries. *Philos. Trans. R. Soc. B Biol. Sci.* **2008**, *364*, 855–859. [CrossRef] [PubMed]
9. Vallortigara, G.; Rogers, L.J. Survival with an asymmetrical brain: Advantages and disadvantages of cerebral lateralization. *Behav. Brain Sci.* **2005**, *28*, 575–589. [CrossRef]
10. Huber, B.A.; Sinclair, B.J.; Schmitt, M. The evolution of asymmetric genitalia in spiders and insects. *Biol. Rev.* **2007**, *82*, 647–698. [CrossRef]
11. Cheng, K. Learning in Cnidaria: A systematic review. *Learn. Behav.* **2021**, 1–15. [CrossRef]
12. Hopkins, W.D.; Misiura, M.; Pope, S.M.; Latash, E.M. Behavioral and brain asymmetries in primates: A preliminary evaluation of two evolutionary hypotheses. *Ann. N. Y. Acad. Sci.* **2015**, *1359*, 65–83. [CrossRef]
13. Henderson, V.W. Paul Broca's Less Heralded Contributions to Aphasia Research. *Arch. Neurol.* **1986**, *43*, 609–612. [CrossRef]
14. de Haan, E.H.; Fabri, M.; Dijkerman, H.C.; Foschi, N.; Lattanzi, S.; Pinto, Y. Unified tactile detection and localisation in split-brain patients. *Cortex* **2020**, *124*, 217–223. [CrossRef]
15. Esteves, M.; Lopes, S.S.; Almeida, A.; Sousa, N.; Leite-Almeida, H. Unmasking the relevance of hemispheric asymmetries—Break on through (to the other side). *Prog. Neurobiol.* **2020**, *192*, 101823. [CrossRef]
16. Francks, C. Exploring human brain lateralization with molecular genetics and genomics. *Ann. N. Y. Acad. Sci.* **2015**. *1359*, 1–13. [CrossRef] [PubMed]
17. Lemée, J.-M.; Bernard, F.; Ter Minassian, A.; Menei, P. Right Hemisphere Cognitive Functions: From Clinical and Anatomical Bases to Brain Mapping During Awake Craniotomy. Part II: Neuropsychological Tasks and Brain Mapping. *World Neurosurg.* **2018**, *118*, 360–367. [CrossRef]
18. Pinto, Y.; De Haan, E.H.; Lamme, V.A. The Split-Brain Phenomenon Revisited: A Single Conscious Agent with Split Perception. *Trends Cogn. Sci.* **2017**, *21*, 835–851. [CrossRef] [PubMed]
19. Pearce, J. The "split brain" and Roger Wolcott Sperry (1913–1994). *Rev. Neurol.* **2019**, *175*, 217–220. [CrossRef] [PubMed]
20. Hopkins, W.D. A review of performance asymmetries in hand skill in nonhuman primates with a special emphasis on chimpanzees. *Chang. Brains Appl. Brain Plast. Adv. Recover Hum. Abil.* **2018**, *238*, 57–89. [CrossRef]
21. Meguerditchian, A.; Vauclair, J.; Hopkins, W.D. On the origins of human handedness and language: A comparative review of hand preferences for bimanual coordinated actions and gestural communication in nonhuman primates. *Dev. Psychobiol.* **2013**, *55*, 637–650. [CrossRef]
22. Ocklenburg, S.; Beste, C.; Arning, L.; Peterburs, J.; Güntürkün, O. The ontogenesis of language lateralization and its relation to handedness. *Neurosci. Biobehav. Rev.* **2014**, *43*, 191–198. [CrossRef] [PubMed]
23. Malatesta, G.; Marzoli, D.; Apicella, F.; Abiuso, C.; Muratori, F.; Forrester, G.S.; Vallortigara, G.; Scattoni, M.L.; Tommasi, L. Received Cradling Bias during the First Year of Life: A Retrospective Study on Children With Typical and Atypical Development. *Front. Psychiatry* **2020**, *11*, 91. [CrossRef] [PubMed]
24. Malatesta, G.; Marzoli, D.; Rapino, M.; Tommasi, L. The left-cradling bias and its relationship with empathy and depression. *Sci. Rep.* **2019**, *9*, 6141. [CrossRef] [PubMed]
25. Malatesta, G.; Marzoli, D.; Tommasi, L. The association between received maternal cradling and neurodevelopment: Is left better? *Med. Hypotheses* **2020**, *134*, 109442. [CrossRef]
26. Gallup, G.G.; Platek, S.M.; Spaulding, K.N. The nature of visual self-recognition revisited. *Trends Cogn. Sci.* **2014**, *18*, 57–58. [CrossRef] [PubMed]
27. Hu, C.; Di, X.; Eickhoff, S.B.; Zhang, M.; Peng, K.; Guo, H.; Sui, J. Distinct and common aspects of physical and psychological self-representation in the brain: A meta-analysis of self-bias in facial and self-referential judgements. *Neurosci. Biobehav. Rev.* **2016**, *61*, 197–207. [CrossRef]
28. Carmody, D.P.; Lewis, M. Self Representation in Children with and without Autism Spectrum Disorders. *Child Psychiatry Hum. Dev.* **2011**, *43*, 227–237. [CrossRef]
29. Lewis, M.; Carmody, D.P. Self-representation and brain development. *Dev. Psychol.* **2008**, *44*, 1329–1334. [CrossRef]
30. Lewis, M.; Sullivan, M.W.; Stanger, C.; Weiss, M.; Sullivan, M.W. Self Development and Self-Conscious Emotions. *Child Dev.* **1989**, *60*, 146–156. [CrossRef]

31. Hecht, E.; Mahovetz, L.M.; Preuss, T.M.; Hopkins, W.D. A neuroanatomical predictor of mirror self-recognition in chimpanzees. *Soc. Cogn. Affect. Neurosci.* **2016**, *12*, 37–48. [CrossRef]
32. Hopkins, W.D.; Latzman, R.D.; Mahovetz, L.M.; Li, X.; Roberts, N. Investigating individual differences in chimpanzee mirror self-recognition and cortical thickness: A vertex-based and region-of-interest analysis. *Cortex* **2019**, *118*, 306–314. [CrossRef]
33. Soler, M.; Colmenero, J.M.; Pérez-Contreras, T.; Peralta-Sánchez, J.M. Replication of the mirror mark test experiment in the magpie (Pica pica) does not provide evidence of self-recognition. *J. Comp. Psychol.* **2020**, *134*, 363–371. [CrossRef]
34. Soler, M.; Pérez-Contreras, T.; Peralta-Sánchez, J.M. Mirror-Mark Tests Performed on Jackdaws Reveal Potential Methodological Problems in the Use of Stickers in Avian Mark-Test Studies. *PLoS ONE* **2014**, *9*, e86193. [CrossRef] [PubMed]
35. Povinelli, D.J. Arboreal Clambering and the Evolution of Self-Conception. *Q. Rev. Biol.* **1995**, *70*, 393–421. [CrossRef]
36. Tresan, D.I. This new science of ours: A more or less systematic history of consciousness and transcendence Part I. *J. Anal. Psychol.* **2004**, *49*, 193–216. [CrossRef]
37. Bassiri, N. Material translations in the Cartesian brain. *Stud. Hist. Philos. Sci. Part C Stud. Hist. Philos. Biol. Biomed. Sci.* **2012**, *43*, 244–255. [CrossRef] [PubMed]
38. Berhouma, M. Beyond the pineal gland assumption: A neuroanatomical appraisal of dualism in Descartes' philosophy. *Clin. Neurol. Neurosurg.* **2013**, *115*, 1661–1670. [CrossRef]
39. López-Muñoz, F.; Marín, F.; Alamo, C. The historical background of the pineal gland: I. From a spiritual valve to the seat of the soul. *Revista de Neurología* **2010**, *50*, 50–57. [PubMed]
40. López-Muñoz, F.; Rubio, G.; Molina, J.; Alamo, C. La glándula pineal como instrumento físico de las facultades del alma: Una conexión histórica persistente. *Neurología* **2012**, *27*, 161–168. [CrossRef]
41. Feinberg, T.E.; Keenan, J.P. Where in the brain is the self? *Conscious. Cogn.* **2005**, *14*, 661–678. [CrossRef] [PubMed]
42. Keenan, J.P.; Falk, D.; Gallup, G.G. *Face in the Mirror*; Harper Collins: New York, NY, USA, 2001.
43. Darwin, C. Sexual Selection in Relation to Monkeys. *Nat. Cell Biol.* **1876**, *15*, 18–19. [CrossRef]
44. Darwin, C. A Biographical Sketch of an Infant. *Ann. Neurosci.* **2010**, *17*, 187–190. [CrossRef]
45. Schmidt, M. *Beobachtungen am Orang-Utan*; Zoologische Garten: Berlin, Germany, 1878; Volume 19.
46. Preyer, W. The Mind of the Child, Part II. In *The Development of the Intellect*; Appleton and Company: New York, NY, USA, 1889.
47. Yerkes, R.M.; Yerkes, A. *The Great Apes*; Yale University Press: New Haven, CT, USA, 1929.
48. Huntley, C.W. Judgments of self based upon records of expressive behavior. *J. Abnorm. Soc. Psychol.* **1940**, *35*, 398–427. [CrossRef]
49. Gesell, A.; Illg, F.L. Infant and Child in the Culture of Today. *Am. Sociol. Rev.* **1943**, *8*, 489. [CrossRef]
50. Hayes, K.J.; Hayes, C. The cultural capacity of chimpanzee. *Hum. Biol.* **1954**, *26*, 288–303. [PubMed]
51. Gallup, G.G. Chimpanzees: Self-Recognition. *Science* **1970**, *167*, 86–87. [CrossRef] [PubMed]
52. Gallup, G.G.; Anderson, J.R. The "olfactory mirror" and other recent attempts to demonstrate self-recognition in non-primate species. *Behav. Process.* **2018**, *148*, 16–19. [CrossRef]
53. Chakraborty, A.; Chakrabarti, B. Is it me? Self-recognition bias across sensory modalities and its relationship to autistic traits. *Mol. Autism* **2015**, *6*, 20. [CrossRef]
54. Chakraborty, A.; Chakrabarti, B. Looking at My Own Face: Visual Processing Strategies in Self–Other Face Recognition. *Front. Psychol.* **2018**, *9*, 121. [CrossRef]
55. Delfour, F.; Marten, K. Mirror image processing in three marine mammal species: Killer whales (*Orcinus orca*), false killer whales (*Pseudorca crassidens*) and California sea lions (*Zalophus californianus*). *Behav. Process.* **2001**, *53*, 181–190. [CrossRef]
56. Morrison, R.; Reiss, D. Precocious development of self-awareness in dolphins. *PLoS ONE* **2018**, *13*, e0189813. [CrossRef]
57. Reiss, D.; Marino, L. Mirror self-recognition in the bottlenose dolphin: A case of cognitive convergence. *Proc. Natl. Acad. Sci. USA* **2001**, *98*, 5937–5942. [CrossRef]
58. Baragli, P.; Demuru, E.; Scopa, C.; Palagi, E. Are horses capable of mirror self-recognition? A pilot study. *PLoS ONE* **2017**, *12*, e0176717. [CrossRef]
59. Hamdan, A.; Ab Latip, M.Q.; Abu Hassim, H.; Noor, M.H.M.; Azizan, T.R.P.T.; Mustapha, N.M.; Ahmad, H. A preliminary study of mirror-induced self-directed behaviour on wildlife at the Royal Belum Rainforest Malaysia. *Sci. Rep.* **2020**, *10*, 1–9. [CrossRef]
60. Plotnik, J.M.; de Waal, F.B.; Moore, D.; Reiss, D. Self-recognition in the Asian elephant and future directions for cognitive research with elephants in zoological settings. *Zoo Biol.* **2010**, *29*, 179–191. [CrossRef]
61. Plotnik, J.M.; de Waal, F.B.M.; Reiss, D. Self-recognition in an Asian elephant. *Proc. Natl. Acad. Sci. USA* **2006**, *103*, 17053–17057. [CrossRef] [PubMed]
62. Brecht, K.F.; Müller, J.; Nieder, A. Carrion crows (Corvus corone corone) fail the mirror mark test yet again. *J. Comp. Psychol.* **2020**, *134*, 372–378. [CrossRef] [PubMed]
63. Kraft, F.-L.; Forštová, T.; Urhan, A.U.; Exnerová, A.; Brodin, A. No evidence for self-recognition in a small passerine, the great tit (Parus major) judged from the mark/mirror test. *Anim. Cogn.* **2017**, *20*, 1049–1057. [CrossRef] [PubMed]
64. Prior, H.; Schwarz, A.; Güntürkün, O. Mirror-Induced Behavior in the Magpie (Pica pica): Evidence of Self-Recognition. *PLoS Biol.* **2008**, *6*, e202. [CrossRef]
65. Uchino, E.; Watanabe, S. Self-recognition in pigeons revisited. *J. Exp. Anal. Behav.* **2014**, *102*, 327–334. [CrossRef] [PubMed]
66. Vanhooland, L.-C.; Bugnyar, T.; Massen, J.J.M. Crows (Corvus corone ssp.) check contingency in a mirror yet fail the mirror-mark test. *J. Comp. Psychol.* **2020**, *134*, 158–169. [CrossRef] [PubMed]
67. Epstein, R.; Lanza, R.; Skinner, B.F. Self-Awareness in the Pigeon. *Science* **1981**, *212*, 695–696. [CrossRef] [PubMed]

68. Anderson, J.R.; Gallup, G.G. Mirror self-recognition: A review and critique of attempts to promote and engineer self-recognition in primates. *Primates* **2015**, *56*, 317–326. [CrossRef]
69. Horowitz, A. Smelling themselves: Dogs investigate their own odours longer when modified in an "olfactory mirror" test. *Behav. Process.* **2017**, *143*, 17–24. [CrossRef]
70. Feinberg, T.E.; Roane, D. Self-representation in delusional misidentification and confabulated "others". *Cortex* **2017**, *87*, 118–128. [CrossRef]
71. Feinberg, T.E.; Roane, D.M. Delusional Misidentification. *Psychiatr. Clin. N. Am.* **2005**, *28*, 665–683. [CrossRef]
72. Feinberg, T.E.; Shapiro, R.M. Misidentification-reduplication and the right hemisphere. *Neuropsychiatry Neuropsychol. Behav. Neurol.* **1989**, *2*, 39–48.
73. Feinberg, T.E.; Venneri, A.; Simone, A.M.; Fan, Y.; Northoff, G. The neuroanatomy of asomatognosia and somatoparaphrenia. *J. Neurol. Neurosurg. Psychiatry* **2009**, *81*, 276–281. [CrossRef]
74. Preilowski, B. Self-recognition as a test of consciousness in the left and right hemisphere of "split-brain" patients. *Act. Nerv. Super.* **1977**, *19* (Suppl. 2), 343–344.
75. Keenan, J.P.; Nelson, A.C.G.; O'Connor, M.; Pascual-Leone, A. Self-recognition and the right hemisphere. *Nat. Cell Biol.* **2001**, *409*, 305. [CrossRef]
76. Kim, M.-K.; Yoon, H.-J.; Shin, Y.-B.; Lee, S.-K.; Kim, J.-J. Neural basis of distorted self-face recognition in social anxiety disorder. *NeuroImage Clin.* **2016**, *12*, 956–964. [CrossRef] [PubMed]
77. Kurth, S.; Moyse, E.; Bahri, M.A.; Salmon, E.; Bastin, C. Recognition of personally familiar faces and functional connectivity in Alzheimer's disease. *Cortex* **2015**, *67*, 59–73. [CrossRef]
78. Morita, T.; Kosaka, H.; Saito, D.N.; Ishitobi, M.; Munesue, T.; Itakura, S.; Omori, M.; Okazawa, H.; Wada, Y.; Sadato, N. Emotional responses associated with self-face processing in individuals with autism spectrum disorders: An fMRI study. *Soc. Neurosci.* **2012**, *7*, 223–239. [CrossRef] [PubMed]
79. Pujol, J.; Giménez, M.; Ortiz, H.; Soriano-Mas, C.; Lopez-Sola, M.; Farre, M.; Deus, J.; Merlo-Pich, E.; Harrison, B.J.; Cardoner, N.; et al. Neural response to the observable self in social anxiety disorder. *Psychol. Med.* **2012**, *43*, 721–731. [CrossRef] [PubMed]
80. Uddin, L.Q. The self in autism: An emerging view from neuroimaging. *Neurocase* **2011**, *17*, 201–208. [CrossRef]
81. Platek, S.M.; Kemp, S.M. Is family special to the brain? An event-related fMRI study of familiar, familial, and self-face recognition. *Neuropsychologia* **2009**, *47*, 849–858. [CrossRef] [PubMed]
82. Kircher, T.T.; Senior, C.; Phillips, M.L.; Benson, P.J.; Bullmore, E.; Brammer, M.; Simmons, A.; Williams, S.C.; Bartels, M.; David, A.S. Towards a functional neuroanatomy of self processing: Effects of faces and words. *Cogn. Brain Res.* **2000**, *10*, 133–144. [CrossRef]
83. Kircher, T.T.; Senior, C.; Phillips, M.L.; Rabe-Hesketh, S.; Benson, P.J.; Bullmore, E.T.; Brammer, M.; Simmons, A.; Bartels, M.; David, A.S. Recognizing one's own face. *Cognition* **2001**, *78*, B1–B15. [CrossRef]
84. Oikawa, H.; Sugiura, M.; Sekiguchi, A.; Tsukiura, T.; Miyauchi, C.M.; Hashimoto, T.; Takano-Yamamoto, T.; Kawashima, R. Self-face evaluation and self-esteem in young females: An fMRI study using contrast effect. *NeuroImage* **2012**, *59*, 3668–3676. [CrossRef]
85. Sugiura, M. Neural basis of self-face recognition: Social aspects. *Brain Nerve* **2012**, *64*, 753–760.
86. Sugiura, M. Three faces of self-face recognition: Potential for a multi-dimensional diagnostic tool. *Neurosci. Res.* **2015**, *90*, 56–64. [CrossRef] [PubMed]
87. Sugiura, M.; Miyauchi, C.M.; Kotozaki, Y.; Akimoto, Y.; Nozawa, T.; Yomogida, Y.; Hanawa, S.; Yamamoto, Y; Sakuma, A.; Nakagawa, S.; et al. Neural Mechanism for Mirrored Self-face Recognition. *Cereb. Cortex* **2014**, *25*, 2806–2814. [CrossRef]
88. Sugiura, M.; Sassa, Y.; Jeong, H.; Horie, K.; Sato, S.; Kawashima, R. Face-specific and domain-general characteristics of cortical responses during self-recognition. *NeuroImage* **2008**, *42*, 414–422. [CrossRef] [PubMed]
89. Sugiura, M.; Sassa, Y.; Jeong, H.; Miura, N.; Akitsuki, Y.; Horie, K.; Sato, S.; Kawashima, R. Multiple brain networks for visual self-recognition with different sensitivity for motion and body part. *NeuroImage* **2006**, *32*, 1905–1917. [CrossRef] [PubMed]
90. Sugiura, M.; Sassa, Y.; Jeong, H.; Wakusawa, K.; Horie, K.; Sato, S.; Kawashima, R. Self-face recognition in social context. *Hum. Brain Mapp.* **2011**, *33*, 1364–1374. [CrossRef] [PubMed]
91. Sugiura, M.; Watanabe, J.; Maeda, Y.; Matsue, Y.; Fukuda, H.; Kawashima, R. Cortical mechanisms of visual self-recognition. *NeuroImage* **2005**, *24*, 143–149. [CrossRef]
92. Morita, T.; Asada, M.; Naito, E. Right-hemispheric Dominance in Self-body Recognition is Altered in Left-handed Individuals. *Neuroscience* **2020**, *425*, 68–89. [CrossRef]
93. Morita, T.; Itakura, S.; Saito, D.N.; Nakashita, S.; Harada, T.; Kochiyama, T.; Sadato, N. The Role of the Right Prefrontal Cortex in Self-evaluation of the Face: A Functional Magnetic Resonance Imaging Study. *J. Cogn. Neurosci.* **2008**, *20*, 342–355. [CrossRef]
94. Morita, T.; Saito, D.N.; Ban, M.; Shimada, K.; Okamoto, Y.; Kosaka, H.; Okazawa, H.; Asada, M.; Naito, E. Self-face recognition shares brain regions active during proprioceptive illusion in the right inferior fronto-parietal superior longitudinal fasciculus III network. *Neuroscience* **2017**, *348*, 288–301. [CrossRef] [PubMed]
95. Morita, T.; Saito, D.N.; Ban, M.; Shimada, K.; Okamoto, Y.; Kosaka, H.; Okazawa, H.; Asada, M.; Naito, E. Self-Face Recognition Begins to Share Active Region in Right Inferior Parietal Lobule with Proprioceptive Illusion During Adolescence. *Cereb. Cortex* **2018**, *28*, 1532–1548. [CrossRef]

96. Morita, T.; Tanabe, H.C.; Sasaki, A.T.; Shimada, K.; Kakigi, R.; Sadato, N. The anterior insular and anterior cingulate cortices in emotional processing for self-face recognition. *Soc. Cogn. Affect. Neurosci.* **2013**, *9*, 570–579. [CrossRef]
97. Platek, S.M.; Wathne, K.; Tierney, N.G.; Thomson, J.W. Neural correlates of self-face recognition: An effect-location meta-analysis. *Brain Res.* **2008**, *1232*, 173–184. [CrossRef]
98. Van Veluw, S.J.; Chance, S.A. Differentiating between self and others: An ALE meta-analysis of fMRI studies of self-recognition and theory of mind. *Brain Imaging Behav.* **2013**, *8*, 24–38. [CrossRef]
99. Weissman-Fogel, I.; Granovsky, Y. The "virtual lesion" approach to transcranial magnetic stimulation: Studying the brain–behavioral relationships in experimental pain. *Pain Rep.* **2019**, *4*, e760. [CrossRef]
100. Ziemann, U. TMS in cognitive neuroscience: Virtual lesion and beyond. *Cortex* **2010**, *46*, 124–127. [CrossRef]
101. Uddin, L.Q.; Molnar-Szakacs, I.; Zaidel, E.; Iacoboni, M. rTMS to the right inferior parietal lobule disrupts self–other discrimination. *Soc. Cogn. Affect. Neurosci.* **2006**, *1*, 65–71. [CrossRef]
102. Kaplan, J.T.; Aziz-Zadeh, L.; Uddin, L.Q.; Iacoboni, M. The self across the senses: An fMRI study of self-face and self-voice recognition. *Soc. Cogn. Affect. Neurosci.* **2008**, *3*, 218–223. [CrossRef] [PubMed]
103. Uddin, L.Q.; Mooshagian, E.; Zaidel, E.; Scheres, A.; Margulies, D.; Kelly, C.; Shehzad, Z.; Adelstein, J.S.; Castellanos, F.; Biswal, B.B.; et al. Residual functional connectivity in the split-brain revealed with resting-state functional MRI. *NeuroReport* **2008**, *19*, 703–709. [CrossRef]
104. Kramer, R.; Duran, K.; Soder, H.; Applegate, L.; Youssef, A.; Criscione, M.; Keenan, J.P. The Special Brain: Subclinical Grandiose Narcissism and Self-Face Recognition in the Right Prefrontal Cortex. *Am. J. Psychol.* **2020**, *133*, 487–500. [CrossRef]
105. Zeugin, D.; Notter, M.P.; Knebel, J.-F.; Ionta, S. Temporo-parietal contribution to the mental representations of self/other face. *Brain Cogn.* **2020**, *143*, 105600. [CrossRef]
106. Payne, S.; Tsakiris, M. Anodal transcranial direct current stimulation of right temporoparietal area inhibits self-recognition. *Cogn. Affect. Behav. Neurosci.* **2017**, *17*, 1–8. [CrossRef] [PubMed]
107. Barrelle, A.; Luauté, J.-P. Capgras Syndrome and Other Delusional Misidentification Syndromes. *Front. Neurol. Neurosci.* **2017**, *42*, 35–43. [CrossRef] [PubMed]
108. Kakegawa, Y.; Isono, O.; Hanada, K.; Nishikawa, T. Incidence and lesions causative of delusional misidentification syndrome after stroke. *Brain Behav.* **2020**, *10*, 01829. [CrossRef]
109. Breen, N.; Caine, D.; Coltheart, M. Mirrored-self Misidentification: Two Cases of Focal Onset Dementia. *Neurocase* **2001**, *7*, 239–254. [CrossRef]
110. Gil-Ruiz, N.; Osorio, R.S.; Cruz, I.; Agüera-Ortiz, L.; Olazarán, J.; Sacks, H.; Álvarez-Linera, J.; Martínez-Martín, P.; Alzheimer Center of the Queen Sofia Foundation; Multidisciplinary Therapy Group. An effective environmental intervention for management of the 'mirror sign' in a case of probable Lewy body dementia. *Neurocase* **2013**, *19*, 1–13. [CrossRef]
111. Mulcare, J.L.; Nicolson, S.E.; Bisen, V.S.; Sostre, S.O. The Mirror Sign: A Reflection of Cognitive Decline? *J. Psychosom. Res.* **2012**, *53*, 188–192. [CrossRef]
112. Phillips, M.L. Mirror, Mirror on the Wall, Who? Towards a Model of Visual Self-recognition. *Cogn. Neuropsychiatry* **1996**, *1*, 153–164. [CrossRef] [PubMed]
113. Villarejo, A.; Puertas-Martin, V.; Moreno-Ramos, T.; Camacho, A.; Porta-Etessam, J.; Bermejo-Pareja, F. Mirrored-self misidentification in a patient without dementia: Evidence for right hemispheric and bifrontal damage. *Neurocase* **2010**, *17*, 276–284. [CrossRef] [PubMed]
114. Yoshida, T.; Yuki, N.; Nakagawa, M. Complex visual hallucination and mirror sign in posterior cortical atrophy. *Acta Psychiatr. Scand.* **2006**, *114*, 62–65. [CrossRef] [PubMed]
115. Roane, D.M.; Feinberg, T.E.; Liberta, T.A. Delusional Misidentification of the Mirror Image. *Curr. Neurol. Neurosci. Rep.* **2019**, *19*, 55. [CrossRef]
116. Bauer, R.M. The Cognitive Psychophysiology of Prosopagnosia. In *Aspects of Face Processing*; Springer Science and Business Media LLC: Berlin/Heidelberg, Germany, 1986; pp. 253–267.
117. Chen, Y.; Slinger, M.; Edgar, J.C.; Bloy, L.; Kuschner, E.S.; Kim, M.; Green, H.L.; Chiang, T.; Yount, T.; Liu, S.; et al. Maturation of hemispheric specialization for face encoding during infancy and toddlerhood. *Dev. Cogn. Neurosci.* **2021**, *48*, 100918. [CrossRef]
118. Gainotti, G. The influence of handedness on hemispheric representation of tools: A survey. *Brain Cogn.* **2015**, *94*, 10–16. [CrossRef]
119. Hewetson, R.; Cornwell, P.; Shum, D.H.K. Relationship and Social Network Change in People with Impaired Social Cognition Post Right Hemisphere Stroke. *Am. J. Speech-Lang. Pathol.* **2021**, *30*, 962–973. [CrossRef] [PubMed]
120. O'Connell, K.; Marsh, A.A.; Edwards, D.F.; Dromerick, A.W.; Seydell-Greenwald, A. Emotion recognition impairments and social well-being following right-hemisphere stroke. *Neuropsychol. Rehabil.* **2021**, *384*, 1–19. [CrossRef] [PubMed]
121. Okubo, M.; Ishikawa, K.; Kobayashi, A. No trust on the left side: Hemifacial asymmetries for trustworthiness and emotional expressions. *Brain Cogn.* **2013**, *82*, 181–186. [CrossRef]
122. Sinha, R.; Dijkshoorn, A.B.C.; Li, C.; Manly, T.; Price, S.J. Glioblastoma surgery related emotion recognition deficits are associated with right cerebral hemisphere tract changes. *Brain Commun.* **2020**, *2*, fcaa169. [CrossRef] [PubMed]
123. Cheng, Y.; Chen, C.; Lin, C.-P.; Chou, K.-H.; Decety, J. Love hurts: An fMRI study. *NeuroImage* **2010**, *51*, 923–929. [CrossRef] [PubMed]
124. Fourie, M.M.; Stein, D.J.; Solms, M.; Gobodo-Madikizela, P.; Decety, J. Empathy and moral emotions in post-apartheid South Africa: An fMRI investigation. *Soc. Cogn. Affect. Neurosci.* **2017**, *12*, 881–892. [CrossRef]

125. Dawson, G.; McKissick, F.C. Self-recognition in autistic children. *J. Autism Dev. Disord.* **1984**, *14*, 383–394. [CrossRef]
126. Dunphy-Lelii, S.; Wellman, H.M. Delayed self-recognition in autism: A unique difficulty? *Res. Autism Spectr. Disord.* **2012**, *6*, 212–223. [CrossRef]
127. Ferrari, M.; Matthews, W.S. Self-recognition deficits in autism: Syndrome-specific or general developmental delay? *J. Autism Dev. Disord.* **1983**, *13*, 317–324. [CrossRef] [PubMed]
128. Mitchell, R.W. A comparison of the self-awareness and kinesthetic-visual matching theories of self-recognition: Autistic children and others. *Ann. N. Y. Acad. Sci.* **1997**, *818*, 39–62. [CrossRef] [PubMed]
129. Neuman, C.J.; Hill, S.D. Self-recognition and stimulus preference in autistic children. *Dev. Psychobiol.* **1978**, *11*, 571–578. [CrossRef] [PubMed]
130. Spiker, D.; Ricks, M. Visual self-recognition in autistic children: Developmental relationships. *Child Dev.* **1984**, *55*, 214–225. [CrossRef]
131. Liu, G.; Zhang, N.; Teoh, J.Y.; Egan, C.; Zeffiro, T.A.; Davidson, R.J.; Quevedo, K. Self-compassion and dorsolateral prefrontal cortex activity during sad self-face recognition in depressed adolescents. *Psychol. Med.* **2020**, 1–10. [CrossRef] [PubMed]
132. Quevedo, K.; Harms, M.; Sauder, M.; Scott, H.; Mohamed, S.; Thomas, K.M.; Schallmo, M.-P.; Smyda, G. The neurobiology of self face recognition among depressed adolescents. *J. Affect. Disord.* **2018**, *229*, 22–31. [CrossRef]
133. Quevedo, K.; Liu, G.; Teoh, J.Y.; Ghosh, S.; Zeffiro, T.; Ahrweiler, N.; Zhang, N.; Wedan, R.; Oh, S.; Guercio, G.; et al. Neurofeedback and neuroplasticity of visual self-processing in depressed and healthy adolescents: A preliminary study. *Dev. Cogn. Neurosci.* **2019**, *40*, 100707. [CrossRef]
134. Quevedo, K.; Ng, R.; Scott, H.; Martin, J.; Smyda, G.; Keener, M.; Oppenheimer, C.W. The neurobiology of self-face recognition in depressed adolescents with low or high suicidality. *J. Abnorm. Psychol.* **2016**, *125*, 1185–1200. [CrossRef]
135. Quevedo, K.; Teoh, J.Y.; Engstrom, M.; Wedan, R.; Santana-Gonzalez, C.; Zewde, B.; Porter, D.; Kadosh, K.C. Amygdala Circuitry During Neurofeedback Training and Symptoms' Change in Adolescents With Varying Depression. *Front. Behav. Neurosci.* **2020**, *14*, 110. [CrossRef]
136. Hirot, F.; Lesage, M.; Pedron, L.; Meyer, I.; Thomas, P.; Cottencin, O.; Guardia, D. Impaired processing of self-face recognition in anorexia nervosa. *Eat. Weight Disord. Stud. Anorex. Bulim. Obes.* **2016**, *21*, 31–40. [CrossRef]
137. Heinisch, C.; Wiens, S.; Grundl, M.; Juckel, G.; Brüne, M. Self-face recognition in schizophrenia is related to insight. *Eur. Arch. Psychiatry Clin. Neurosci.* **2013**, *263*, 655–662. [CrossRef] [PubMed]
138. Barone, P.; Corradi, G.; Gomila, A. Infants' performance in spontaneous-response false belief tasks: A review and meta-analysis. *Infant Behav. Dev.* **2019**, *57*, 101350. [CrossRef] [PubMed]
139. Frank, C.K. Reviving pragmatic theory of theory of mind. *AIMS Neurosci.* **2018**, *5*, 116–131. [CrossRef]
140. Tomasello, M. How children come to understand false beliefs: A shared intentionality account. *Proc. Natl. Acad. Sci. USA* **2018**, *115*, 8491–8498. [CrossRef]
141. Krych-Appelbaum, M.; Law, J.B.; Jones, D.; Barnacz, A.; Johnson, A.; Keenan, J.P. "I think I know what you mean": The role of theory of mind in collaborative communication. *Interact. Stud.* **2007**, *8*, 267–280. [CrossRef]
142. Feinberg, T. *Altered Egos: How the Brain Creates the Self*; Oxford University Press: Oxford, UK, 2002.

MDPI
St. Alban-Anlage 66
4052 Basel
Switzerland
Tel. +41 61 683 77 34
Fax +41 61 302 89 18
www.mdpi.com

Symmetry Editorial Office
E-mail: symmetry@mdpi.com
www.mdpi.com/journal/symmetry

www.ingramcontent.com/pod-product-compliance
Lightning Source LLC
Chambersburg PA
CBHW062107220926
51850CB00010B/2919
9783036574783